EAI/Springer Innovations in Communication and Computing

Series editor
Imrich Chlamtac, European Alliance for Innovation, Ghent, Belgium

Editor's Note

The impact of information technologies is creating a new world yet not fully understood. The extent and speed of economic, life style and social changes already perceived in everyday life is hard to estimate without understanding the technological driving forces behind it. This series presents contributed volumes featuring the latest research and development in the various information engineering technologies that play a key role in this process.

The range of topics, focusing primarily on communications and computing engineering include, but are not limited to, wireless networks; mobile communication; design and learning; gaming; interaction; e-health and pervasive healthcare; energy management; smart grids; internet of things; cognitive radio networks; computation; cloud computing; ubiquitous connectivity, and in mode general smart living, smart cities, Internet of Things and more. The series publishes a combination of expanded papers selected from hosted and sponsored European Alliance for Innovation (EAI) conferences that present cutting edge, global research as well as provide new perspectives on traditional related engineering fields. This content, complemented with open calls for contribution of book titles and individual chapters, together maintain Springer's and EAI's high standards of academic excellence. The audience for the books consists of researchers, industry professionals, advanced level students as well as practitioners in related fields of activity include information and communication specialists, security experts, economists, urban planners, doctors, and in general representatives in all those walks of life affected ad contributing to the information revolution.

About EAI

EAI is a grassroots member organization initiated through cooperation between businesses, public, private and government organizations to address the global challenges of Europe's future competitiveness and link the European Research community with its counterparts around the globe. EAI reaches out to hundreds of thousands of individual subscribers on all continents and collaborates with an institutional member base including Fortune 500 companies, government organizations, and educational institutions, provide a free research and innovation platform.

Through its open free membership model EAI promotes a new research and innovation culture based on collaboration, connectivity and recognition of excellence by community.

More information about this series at http://www.springer.com/series/15427

Shishir K. Shandilya • Neal Wagner
Atulya K. Nagar
Editors

Advances in Cyber Security Analytics and Decision Systems

Springer

EAI
RESEARCH MEETS INNOVATION

Editors
Shishir K. Shandilya
School of Computing Science
& Engineering
Vellore Institute of Technology
VIT Bhopal University
Bhopal, Madhya Pradesh, India

Neal Wagner
Analytics and Intelligence Division
Systems and Technology Research
Woburn, MA, USA

Atulya K. Nagar
School of Mathematics, Computer Science
and Engineering
Faculty of Science
Liverpool Hope University
Liverpool, UK

ISSN 2522-8595 ISSN 2522-8609 (electronic)
EAI/Springer Innovations in Communication and Computing
ISBN 978-3-030-19352-2 ISBN 978-3-030-19353-9 (eBook)
https://doi.org/10.1007/978-3-030-19353-9

This Springer imprint is published by the registered company Springer Nature Switzerland AG
The registered company address is: Gewerbestrasse 11, 6330 Cham, Switzerland

To my lifelines "Smita," "Samarth," and "Nityaa"

—Shishir K. Shandilya

To my wonderful family, Rahma, Jensine, and Jamal

—Neal Wagner

To my lovely daughters, "Kopal" and "Priyel" and my wife "Jyoti"

—Atulya K. Nagar

Preface

Today's world is witnessing a constant barrage of cyberattacks in the form of ransomware, phishing, malware, botnets, insider threat, and many others. The situation is untenable and getting worse day by day. The amount of data at risk is enormous and rapidly growing over time. Cyber adversaries are becoming more advanced, often utilizing intelligent algorithms and technologies to steal confidential data, disrupt critical networks, and corrupt communications. Therefore, the book's focus is on cybersecurity defensive measures and risk mitigations to counter these ever-growing attacks and make the digital world safer. Cybersecurity consists of the set of methods that seek to provide protection and risk reduction against cyberattacks and maintain network integrity. This subfield of IT security is predicted to quickly grow to over 10% of the total IT market share by the year 2025, which is currently estimated to be a $4 trillion industry. This book is an important cybersecurity analytics resource as it includes the latest, cutting-edge techniques and solutions on the subject.

The book is focused on state-of-the-art methods and algorithms and highlights empirical results along with theoretical concepts to provide a good comprehensive reference for students, researchers, scholars, professionals, and practitioners in the field of cybersecurity and analytics. It provides insight into the practical application of cybersecurity methods so that the readers can understand how abstract ideas can be employed to solve real-world security problems.

The book brings together leading researchers and practitioners in the field and will be an important resource for cybersecurity students, with an aim to promote, present, analyze, and discuss the latest research of the field.

We express our heartfelt gratitude to all the authors, reviewers, and publishers, especially to Eliska Vlckova and Ms. Lucia Zatkova for their kind support. We hope that this book will be beneficial to all the concerned readers.

Bhopal, India — Shishir K. Shandilya
Woburn, MA, USA — Neal Wagner
Liverpool, UK — Atulya K. Nagar

Contents

About the Editors

Shishir K. Shandilya is Division Head of Cybersecurity and Digital Forensics at VIT Bhopal University, India, and Visiting Researcher at Liverpool Hope University, United Kingdom. He is Cambridge University Certified Professional Teacher and Trainer and Senior Member of IEEE-USA. He is a renowned Academician and active Researcher with proven record of teaching and research. He has received "IDA Teaching Excellence Award" for distinctive use of technology in teaching by India Didactics Association, Bangalore (2016), and "Young Scientist Award" for 2 consecutive years (2005 and 2006) by Indian Science Congress and MP Council of Science and Technology. He has written seven books of international fame (published in the USA, Denmark, and India) and published quality research papers. He is an active Member of various international professional bodies.

Neal Wagner is a Complex Systems Scientist at Systems & Technology Research, Massachusetts, USA. His focus lies in developing problem-solving methods, tools, and techniques that combine computational intelligence and modeling and simulation to create automated/semiautomated cyber decision-making systems. Prior to joining Systems & Technology Research, he was a Technical Staff Member of MIT Lincoln Laboratory in the Cyber Analytics and Decision Systems Group where he focused on AI applications to cybersecurity. Prior to MIT, he was at SolveIT Software, where he specialized in the commercialization of bio-inspired computing techniques for supply chain optimization of large organizations. His academic experience includes stints as a Faculty Member of the Computer Science and Information Systems Departments at Augusta University and Fayetteville State University. He holds a BA degree in Mathematics from the University of North Carolina at Asheville and an MS degree in Computer Science and a PhD in Information Technology both from the University of North Carolina at Charlotte.

Atulya K. Nagar holds the Foundation Chair as Professor of Mathematical Sciences and is the Pro-Vice-Chancellor for Research; and Dean of the Faculty of Science at Liverpool Hope University, United Kingdom. He has been the Head of the School of Mathematics, Computer Science and Engineering which he established at the

University. He is an internationally respected Scholar working at the cutting edge of theoretical computer science, applied mathematical analysis, operations research, and systems engineering. He received a prestigious Commonwealth Fellowship for pursuing his doctorate (DPhil) in Applied Nonlinear Mathematics, which he earned from the University of York (UK) in 1996. He holds BSc (Hons), MSc, and MPhil (with distinction) in Mathematical Physics from the MDS University of Ajmer, India. His research expertise spans both applied mathematics and computational methods for nonlinear, complex, and intractable problems arising in science, engineering, and industry. In problems like these, the effect is known, but the cause is not. In this approach of mathematics, also known as "inverse problems," sophisticated mathematical modeling and computational algorithms are required to understand such behavior.

Adaptive Attacker Strategy Development Against Moving Target Cyber Defenses

M. L. Winterrose, K. M. Carter, N. Wagner, and W. W. Streilein

1 Introduction

Today cyber defenders are at a systematic disadvantage in cyber conflict. Attackers often only need to exploit a single security vulnerability to succeed with an attack, and attackers can typically act at a time and place of their choosing. Furthermore, the technological monocultures that dominate information technology today place these systems at significant risk for attack. With a large number of organizations and individuals using essentially identical hardware, operating systems, and application software, significant incentives have been created for cyber attackers to discover and exploit vulnerabilities in these systems.

In this context, new techniques are under development by the cyber security research community to rebalance the playing field for cyber defenders. A major effort in recent years along these lines is an attempt by cyber defenders to diversify the most vulnerable pieces of the existing large cyber monocultures. These techniques, aiming to increase the diversity of a system's attack surface, causing increased operational costs and uncertainty for attackers have come to be grouped under the umbrella term *moving target*. Moving target techniques have been applied

This work is sponsored by the Department of Defense under Air Force Contract FA8721-05-C-0002. Opinions, interpretations, conclusions and recommendations are those of the author and are not necessarily endorsed by the United States Government.

M. L. Winterrose · K. M. Carter · W. W. Streilein
MIT Lincoln Laboratory, Lexington, MA, USA
e-mail: michael.winterrose@ll.mit.edu; kevin.carter@ll.mit.edu; wws@ll.mit.edu

N. Wagner (✉)
Analytics and Intelligence Division, Systems and Technology Research, Woburn, MA, USA
e-mail: neal.wagner@stresearch.com

S. K. Shandilya et al. (eds.), *Advances in Cyber Security Analytics and Decision Systems*, EAI/Springer Innovations in Communication and Computing,
https://doi.org/10.1007/978-3-030-19353-9_1

to diversify runtime environments, software, networks, platforms, and data in recent years (Okhravi et al. 2013, 2014).

In this study, we examine a class of *migration-based* techniques that dynamically change the platform (i.e., operating system [OS]) that is active on a host in order to reduce attacker success and increase attacker resource investment requirements. These techniques work under the assumption that the attacker has limited resources and generally does not have exploits available for *all* OSes. As such, migrating between OSes with some frequency reduces the ability of an attacker to maintain persistence on a system. Additionally, it increases the uncertainty for an attacker that aims to expend resources toward exploit development.

Two recent studies have examined the optimal scheduling policy for a temporal migration moving target defense. In the first, the conclusion was drawn that a uniform random scheduling policy by a defender employing a set of active spam filters performed optimally against an adaptive adversary (Colbaugh and Glass 2012). The second set of studies (Carter et al. 2014) required the attacker to maintain persistence in a system for a period of time before reward accrued. An additional factor incorporated in the second set of studies was that of coupled exploits, in which a given exploit targeted at a specific OS may work against other similar OSes. It was shown that a deterministic scheduling policy that maximizes the diversity of the platforms played in each successive round was superior under the assumption of coupled exploits and the requirement of attacker persistence (Carter et al. 2014).

The goal of this work is to evaluate different scheduling policy strategies against a nondeterministic, adaptive attacker. As opposed to (Colbaugh and Glass 2012) and (Carter et al. 2014), which posited a restrictive attacker model in which an attacker could only develop exploits for OSes presented to it by the defender, we extend our prior work (Winterrose and Carter 2014) to model an adaptive attacker with a less restrictive attacker model. This more flexible adversary model allows an attacker to invest in the development of zero-day exploits against *any* potential defender system. Attackers observe defender actions and use these observations to learn optimal investment strategies. We will demonstrate the complex, yet intuitive, strategies that are evolved to optimize attacker success against various defender platform migration scheduling policies.

The major contributions of this work are as follows:

1. We employ a less restrictive attacker model that more accurately captures the resource investment decision problem faced by an attacker.
2. We present the first use of a novel finite state machine (FSM) construct that transitions between action states based on a heterogeneous set of system observations.
3. We show that learned attacker strategies are highly sensitive to the statistical characteristics of the defender's moving target scheduling policy.
4. We demonstrate that the degree to which a defense policy is optimal against an adaptive adversary changes as the duration of conflict varies.

2 Methods

2.1 Attacker–Defender Game Scenario

Many security scenarios can be modeled as games (Tambe 2012). Typically this involves the reduction of a given adversarial situation to its most essential elements, casting the salient features of a security conflict in stark relief. Done well, this procedure facilitates the discovery of the deeper mechanisms underlying a real-world phenomenon by eliminating the nonessential aspects of a situation.

In our scenario, the attacker, characterized by a population of N strategies, plays a series of games made up of sets of consecutive matches against the defender. A simulated game between the attacker and the defender is played in which time advances in discrete steps. A single match is executed at each tick of the simulation clock. In each match, a deterministic defender activates one platform according to a pre-determined defense strategy. The attacker observes the defender's choice and must decide how to allocate resources in the next round to bring exploits into existence so as to attack the defender's system with an optimal chance of success. All exploits in this study are assumed to be *zero-day* exploits, meaning that they are unobserved by the defender when used against the corresponding operating system.

In the simulated game, there exists one possible zero-day exploit for each type of platform the defender might deploy in the temporal platform migration defensive system. In each match of the game, the attacker may choose to use its resources to further develop one of these zero-day exploits.

The attacker resources (i.e., the number of rounds of attacker resource investment) required to bring a given zero-day exploit into existence is determined by sampling from a Gamma distribution at the beginning of each generation. The attacker is not informed *a priori* of the number of resources that will be required to bring a zero-day exploit against a given platform into existence. Instead, the attacker discovers this only after having successfully created the exploit through the allocation of sufficient resources. Additionally, the attacker must learn when to discontinue investment in the creation of a particular exploit once the required number of resources has been invested, as continued investment will be wasted. This is used to model the fact that a real-world attacker is not generally able to predict *a priori* the level of effort that will be required to develop new exploits against a given system.

In each match, the attacker uses any exploits that have been developed in the current game against the activated platform. Success for the attacker in a match occurs when it has an available exploit that works against the platform activated by the defender. Intuitively, the attacker gains a reward if the attacker is able to compromise the defender's system during the match, and earns nothing otherwise. For the purposes of this study, we impose a *persistence* requirement on the adversary, such that a reward is granted only after three consecutive successful matches. One may view this as the requisite length of time to stage an attack, such as the exfiltration of data over a difficult channel, with a cumulative reward being granted each match while the full attack is successful (i.e., ≥ 3 matches) (Carter et al. 2014).

In this study we allow the defender access to the pool of five platforms: Fedora on ×86, Gentoo on ×86, Debian on ×86_64, FreeBSD on ×86, and CentOS on ×86. In the randomization defense, in each match, an OS is selected from the pool of five for activation on the defender's system uniformly at random, with the caveat that the OS activated in the present match cannot be activated again in the immediate next match. The diversity defense consists of the deterministic activation of Fedora, Debian, and FreeBSD in succession (Carter et al. 2014). This rotation between three platforms maximizes the diversity of the source code presented to the attacker from match to match and reduces the likelihood that an exploit developed for one OS can persist when that OS is replaced by the next OS in the rotation.

The attacker can develop a targeted exploit for each of the defender operating systems. A *targeted exploit* works with certainty each time it is used against the platform it targets. In our model, a developed exploit also works against platforms other than the exploit's target system with a probability proportional to the code similarity of the two operating systems. We term this effectiveness of an exploit against systems other than the target system its *cross-platform* effectiveness.

Table 1 lists a set of *code similarity scores* for the defender's operating systems. These similarity scores were calculated using the *measure of software similarity* (MOSS) tool (Schleimer et al. 2003) and based on each operating system's kernel code and standard device drivers (Carter et al. 2014). The similarity scores are given on a scale from 0 to 1, with 1 implying identical operating system code and 0 indicating completely dissimilar operating system code.

We note that FreeBSD is an outlier in the set with a markedly low similarity score compared with the remainder of the set. This is explained by the fact that FreeBSD is based on Unix while the other four operating systems are Linux-based. In the sections that follow the outlier status of the FreeBSD platform will be shown to have a significant impact on attacker strategy development.

In this study, the cross-platform effectiveness is determined on a match-by-match basis. On each match, the attacker plays all available exploits against the defender activated platform. If the attacker has developed the exploit *targeted* at the activated defender's system, then the attacker succeeds with the attack with certainty. On the other hand, if the targeted exploit has not been created by the attacker for the activated platform, any other created exploits succeed against the defender's system with a probability equal to the similarity score between the exploit's targeted system and the system activated by the defender.

Table 1 Platform similarity scores based on operating system kernel code and standard device drivers

	CentOS	**Fedora**	**Debian**	**Gentoo**	**FreeBSD**
CentOS	1.0000	0.6645	0.8067	0.6973	0.0368
Fedora	0.6645	1.0000	0.5928	0.8658	0.0324
Debian	0.8067	0.5928	1.0000	0.6202	0.0385
Gentoo	0.6973	0.8658	0.6202	1.0000	0.0330
FreeBSD	0.0368	0.0324	0.0385	0.0330	1.0000

Reproduced from Carter et al. (2014)

2.2 *Finite State Machine Strategy Encoding*

We represent attacker strategies as binary chromosomes encoding a FSM construct. An FSM is an abstract machine that takes discrete inputs from an environment and specifies a discrete output in response. An agent modeled by an FSM will occupy only one state at any point in time. Such an agent transitions between states based on observations of its environment.

Each strategy in our 30-strategy population is represented by one 16-state, 160-transition FSM. Each state encodes up to eight possible actions, leading to a 692-bit chromosome encoded in a manner similar to (Miller 1996; Winterrose and Carter 2014). Figure 1 depicts a single state in our machine and its outgoing transitions.

The 30 machines are initialized randomly before the simulation begins. During the simulation, the actions encoded in each state of each machine and the transitions between states evolve according to the genetic algorithm presented in the next section.

We use 16-state FSMs for historical reasons (Miller 1996), but find through ancillary studies that the actual strategies evolved by attackers generally fit easily within our 16-state constraint. An extended study with a widely varying number of

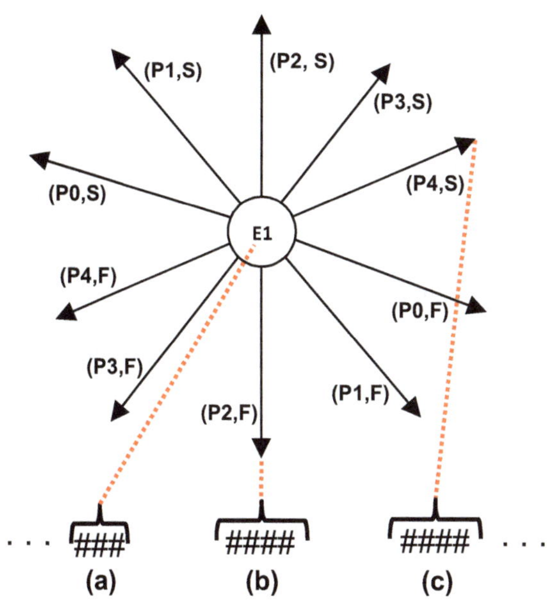

Fig. 1 Hypothetical single automata state and corresponding outgoing transition set of an attacker's FSM (upper). *E*, *P*, *S*, and *F* represent *exploit investment*, *platform observation*, *successful attack*, and *failed attack* by the attacker, respectively. The finite automata state maps into a binary chromosome in which bits are represented by $\# \in \{0, 1\}$. Portion (**a**) of the chromosome encodes the attacker investment in zero-day exploit creation when the attacker occupies this hypothetical state using 3-bits. Segments (**b**) and (**c**) of the chromosome encode two possible transitions executed in response to observation of the defender's activations and the attacker's success in the current round. The transitions are encoded using 4-bits each

machine states would shed useful light on the consequences of *bounded rationality* on the nature of strategies evolved in the cyber domain.

Our FSMs transition between states based on both the type of platform activated by the defender in the previous round and on the success the attacker had with its exploit attacks in the previous round. To the best of our knowledge, this *dual-observation* transition model is unique to this study. Previous studies using a simpler, but related, FSM construct to play the Prisoner's Dilemma game-theoretic scenario (Miller 1996) transitioned between machine states based on a single observation of opponent action in each round of play.

2.3 Evolutionary Algorithm

The adaptive attacker in our study evolves strategies against the defender using a genetic algorithm (GA) (Holland 1975). Originally conceived as a stylized model of biological evolution, the GA has proven to be a robust method that can efficiently search solution spaces that are nonlinear and/or discontinuous. In our implementation, we randomly initialize 30 strategies at the outset of a simulation run. In each generation each *agent* (i.e., *strategy*) plays a *game* consisting of M matches against the defender.

In a *match*, if the defender's active platform is vulnerable to an exploit that has been successfully developed, the attacker accrues a reward, governed by some underlying function that is hidden from the algorithm. This may include immediate reward, or for example, require some level of consecutive success before a reward is granted (e.g., *persistence*). See (Carter et al. 2014) for example scenarios and associated reward functions.

Once the reward is computed, the match is concluded. A new match begins with the attacker choosing an exploit to develop with its allocated resources (one resource is available for investment by the attacker in each match). Concurrently, the defender selects a platform to make active in the system, against which the attacker moves with any available exploits. This continues for M matches, at which point the *game* ends between the chosen attacker and the defender, and a new attacker strategy from the population is rotated in to play against the defender. Once all attackers have played their M matches in the generation (g) against the defender, each strategy, i, is assigned a fitness score, F, based on its success against the defender,

$$F_{i,g} = \sum_{j=1}^{M_g} \Phi_{i,j}, \tag{1}$$

with $\Phi_{i,j}$ set equal to +1 if the system is compromised, and set equal to 0 otherwise. Attacker strategies are ranked based on their fitness scores. A new population of attacker strategies is generated for play against the defender using the following steps:

1. A fraction of the top-ranked attacker strategies are copied directly into the new population. This procedure is known as *elitism* and is commonly used in GA applications to avoid the loss of the best strategies from previous strategy populations (Mitchell 1996).
2. Two attacker strategies are chosen from the current population using fitness proportionate selection in which higher-ranking strategies are more likely to be selected.
3. The two selected (parent) strategies undergo the *crossover* genetic operation (analogous to biological sexual reproduction) to generate two offspring strategies. In this operation, a single crossover point $c \in \{1,2, \ldots n\}$ on each of two parent chromosomes is selected uniformly at random. The first offspring combines the first c bits from the first parent with all bits after the c+1 chromosome position of the second parent to form a new chromosome. The second offspring takes all bits after the c+1 chromosome position from parent 1 and combines it with the first c bits of parent two's chromosome to form a new strategy.
4. The two offspring strategies are then subject to the *mutation* genetic operation (analogous to asexual reproduction). In this operation bits in the chromosome are randomly altered. The mutation operation is commonly used in GA applications to increase population diversity and avoid local extrema in the search space (Michalewicz 1996).
5. The two generated offspring strategies are then added to the new population.
6. The above steps are repeated until the new population has a sufficient number of strategies (specified by a population number parameter).

Attacker strategies are evolved over a set of generations where each generation includes the attacker–defender simulated games and the above steps to generate new populations of strategies.

3 Experiments

3.1 Simulation Initialization

The defender's dynamic platform scheduling policy is assigned at the beginning of the game and is not altered as the game progresses. The attacker's strategy is represented by a population of randomly initialized strategies encoded as binary chromosomes representing FSMs. Each iteration of the simulation is allowed to run for 100 generations of genetic algorithm evolution, with the attacker strategy being evolved in each of these generations. A single generation consists of each of the $N = 30$ attackers playing M matches against the defender. The 100-generation run is iterated 100 times and the results aggregated and averaged to account for the stochasticity in the model.

The number of attacker resources required to bring a given zero-day exploit into existence is determined at the beginning of each generation by independent draws from a Gamma distribution for each of the 5 possible zero-day exploits available for development by the attacker, similar to the procedure we used in (Winterrose and Carter 2014). The Gamma distribution is parameterized by a mean (μ) and variance (σ^2) parameter. We use μ =25 and σ^2 = 10 throughout this study.

In the analysis that follows, we typically extract the fittest learned attacker strategy in each generation of each simulation run and aggregate these together to produce the results discussed. This procedure is consistent with the focus in this paper on the nature of the optimal attacker strategies developed against the defender's moving target defense. We refer to this set of fittest strategies extracted from each simulation run as the *fittest attackers* or *fittest strategies* hereafter.

Simulations were created and executed in the NetLogo modeling environment (Wilensky 1999). Data aggregation across simulation runs and the calculation of statistical measures were carried out using MATLAB release 2013b (Matlab 2013). To visualize the evolved FSMs, we have used the Gephi network visualization and analysis software package (Bastian et al. 2009).

3.2 Attacker's Response to Diversity and Randomization Defense

Figure 2 shows the match-level response at the 100th generation of genetic algorithm evolution for the fittest attackers averaged over 100 simulation runs. The attacker performs better against the randomization defense throughout the 100-match game, but the difference in performance narrows as the match number increases. We recall that the attacker must compromise the defender's system for three consecutive matches using its developed exploits before beginning to accrue a reward for system compromise. Figure 2 shows that this begins to occur at an earlier point in match play when the attacker faces the randomization defense. Specifically, the attacker's fitness begins to rise around match number 40 when the attacker faces the randomization defense, roughly the point at which an efficient attacker might begin to have access to two exploits given this studies' exploit creation cost parameterization. With two exploits created the attacker can utilize the cross-platform effectiveness of the created exploits to achieve the persistence required for accruing attacker reward.

Against the diversity defense, on the other hand, the attacker does not begin accumulating reward until just before match 60. Between the 60th and 75th match the attacker's reward (i.e., fitness) climbs slowly, then accelerates sharply after approximately match 75. This can be understood by recalling that once the attacker has had the opportunity to develop three targeted exploits, it is able to completely counter the diversity defense. This causes the fitness of the attacker facing the diversity defense to quickly approach the fitness of the attacker facing the randomization defense.

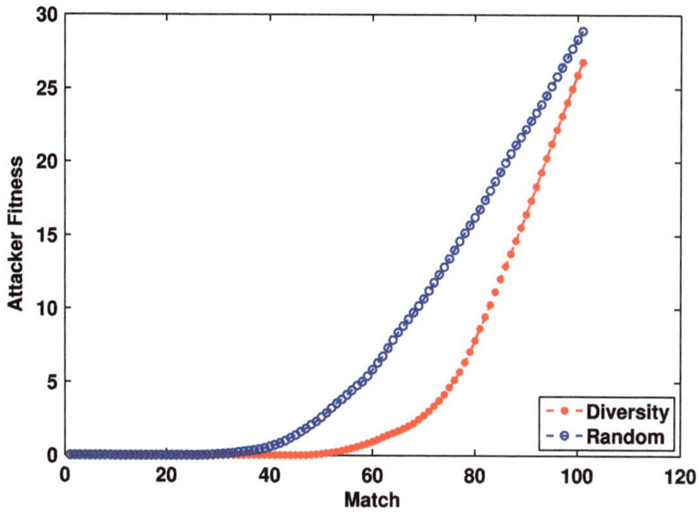

Fig. 2 Fittest attacker game success in the 100th generation of genetic algorithm evolution averaged over 100 simulation runs. The attacker is most successful against the defender deploying the randomization dynamic platform scheduling policy, though the difference in response narrows in later matches. See text for discussion

Figure 3 shows the structural properties of an exemplar FSM encoding an attacker's evolved strategy when facing the diversity defense. In the figure the node and label sizes are proportional to the number of transitions into a given state. The importance of the Fedora, Debian, and FreeBSD exploit development in the learned attacker strategy are clear in this FSM representation. In particular, FreeBSD is the most prevalent investment state in the structure, a fact we discuss in the next section.

3.3 Patterns of Attacker Investment in Zero-Day Exploit Creation

An important consideration when deciding upon a deployment strategy for a dynamic platform moving target defense is how the attacker is likely to alter its strategy based on the defender's choices. For this experiment, we were interested in understanding how the statistical characteristics of the defender's scheduling policy affect attacker exploit creation investment choices.

The basic choice the attacker faces is the manner in which to invest its resource in each round to compromise the attacker's system with maximum effectiveness. The key considerations for the attacker in achieving this goal is the persistence requirement (i.e., three consecutive successful attacks before attacker reward accrues) and the cross-platform effectiveness of each zero-day exploit. The need to weigh these factors together with the observations of the defender's dynamic platform scheduling policy make the investment choice a complex one for the attacker.

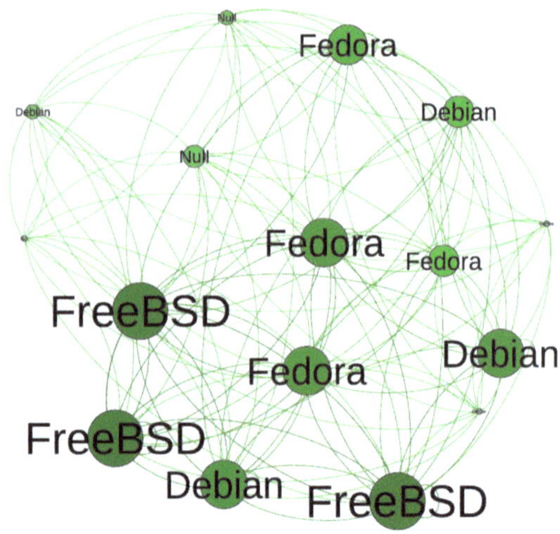

Fig. 3 Structural representation of an exemplar attacker strategy developed to counter the diversity defense. The strategies are encoded in a FSM. Nodes are labeled by the investment to be made by the attacker in the various machine states. Edges represent transitions between states based on observations of the defender's actions and successful game play. See text for further discussion

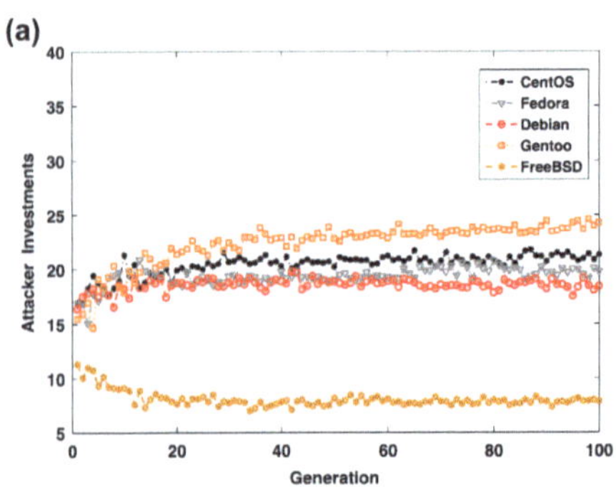

Fig. 4 Generational progression of fittest attacker exploit creation investments averaged over 100 simulation runs. Investment patterns of attackers facing a randomized defender scheduling policy (**a**) differ markedly from the investment patterns developed by the evolving attackers facing a diversity defense (**b**)

Figure 4 shows the generational investment patterns learned by the fittest attackers aggregated across the 100 simulation runs. It is clear that the statistical character of the defender's scheduling policy strongly affects the exploit investment pattern of the attacker. The largest effects are observed in the preference or disdain the attacker shows for developing the FreeBSD zero-day exploit. Figure 4 a shows that when facing the randomizing defender the attacker prefers to minimize investment in the FreeBSD exploit and focus investment on the creation of exploits for the Linux-based platforms. This behavior contrasts sharply with the attacker's response to the diversity-maximizing defender (Fig. 4b). Here the attacker shows a strong preference for developing the FreeBSD exploit.

We note that in Fig. 4 the attacker has discovered these investment patterns already in the initial generation. This early discovery of the fittest attacker strategy is essentially a matter of luck. The process of learning these investment patterns is more evident when examining the evolution of investment patterns within the entire population of $N = 30$ attackers, as shown in Fig. 5. Here we see the population mean investment in each of the available exploits distributed approximately uniformly in the initial generation, then diverging strongly in just a few generations as the population converges on the strategies of avoiding FreeBSD exploit investment when facing the randomized defense (Fig. 5a), and investing heavily in FreeBSD exploit creation when facing the diversity defense in Fig. 5b.

These trends can be understood by taking account of the following observations. When facing the diversity defense the attacker can predict with certainty that it will face a defender activating the FreeBSD platform reliably every 3 matches. Given the dissimilarity of FreeBSD with the other 4 platforms, this makes it improbable that the attacker will achieve the requirement of 3-match persistence across the FreeBSD activation if the FreeBSD exploit has not been created. When the attacker is facing the randomization defense, in contrast, there exists a reasonable probability that the attacker will achieve the persistence requirement and accrue reward without facing activation of the FreeBSD platform by the defender. The predictability of needing to overcome a FreeBSD activation in the first case (i.e., diversity defense), and the uncertainty of facing a FreeBSD activation in the second case (i.e., randomization defense) rationalizes the investment patterns in Figs. 4 and 5.

3.4 Engagement Duration Effects

Another important consideration when deciding upon a deployment strategy for a dynamic platform moving target defense is the duration of the interaction. For this experiment, we were interested in understanding how attacker fitness in the face of diversity and random strategies was affected by different interaction (i.e., game)

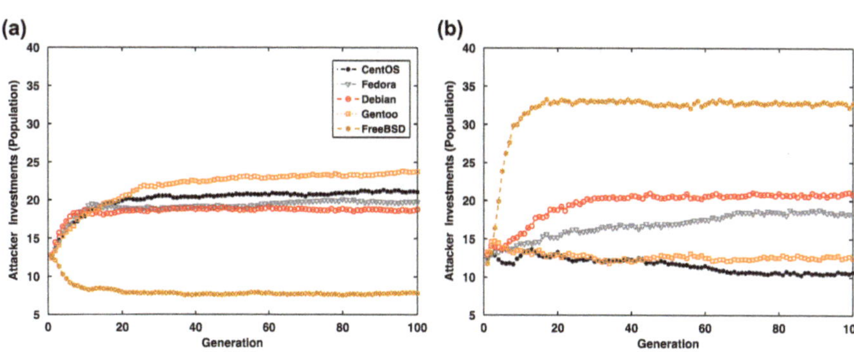

Fig. 5 Generational progression of mean attacker exploit creation investments for the entire attacker population of $N = 30$ strategies. Displayed results are aggregated and averaged over 100 simulation runs. As in the fittest attacker case, investment patterns of attackers facing a randomized defender scheduling policy (**a**) are quite different from the investment patterns developed by the evolving attackers facing a diversity defense (**b**)

durations. Figure 6 demonstrates that the duration of attacker–defender interaction greatly affects the efficacy of the defensive capability, which is reflected inversely in the fitness level of the attacker: better attacker fitness implies worse defensive capability. In Fig. 6a we see that the attacker achieves a high level of fitness for a 75-Match game when a random strategy is utilized. By contrast, when the defender utilizes the diversity strategy, the attacker never achieves a similar level of fitness, though the overall level of fitness does increase with generation. As the duration of the games increases, through 100-Match and 125-Match games, Fig. 6 shows that the value of the diversity strategy diminishes for longer duration interactions. Specifically, when 100-Match games are played, the attacker fitness level is roughly equivalent for random and diversity strategy at the start, with the diversity strategy initially performing suboptimally and then improving. When the 125-match games are played, the diversity strategy is always suboptimal to the random strategy in providing effective defense by allowing the attacker fitness to reach a higher level. It is worth noting that the absolute fitness level achieved by the attacker increases overall as the interaction duration increases regardless of the defensive strategy employed. This is due to the fact that the attacker is provided with more time to develop an exploit in all cases regardless of the defensive strategy in use and thus is able to improve fitness level.

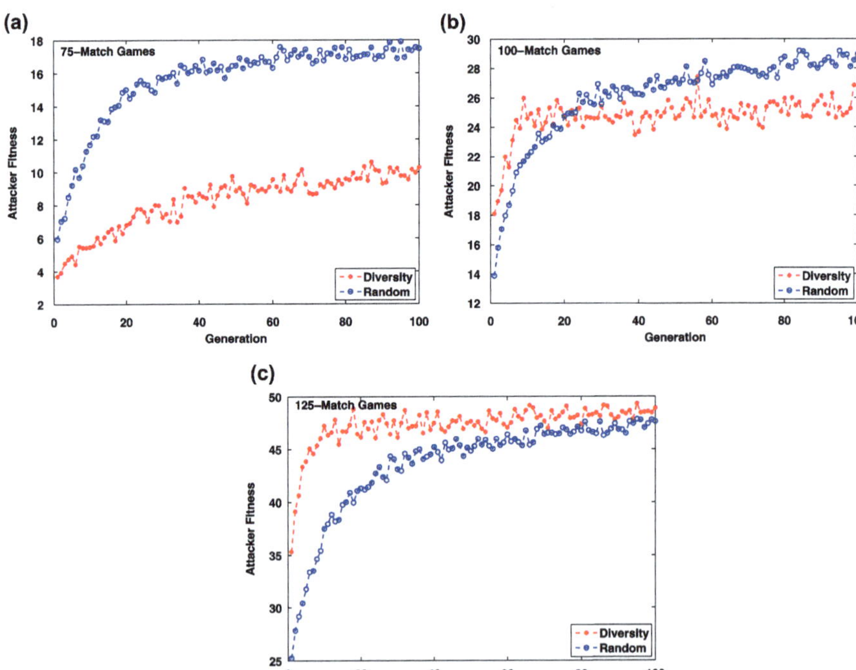

Fig. 6 Game success as a function of generation number for the set of fittest attackers facing the defender in games of varying length, as indicated in the figures

To understand the benefit provided at shorter durations by the diversity strategy it is instructive to consider that although the attacker is able to focus his resources on a smaller set of target OSes, the shorter duration makes it difficult to achieve all exploits needed to compromise the system with the required persistence. As the duration increases, the attacker is more likely to develop targeted exploits for the entire set of the OSes in the diversity strategy before the interaction ends and thus is able to achieve persistent compromise. When the game duration reaches 125 matches, the random strategy provides more defensive benefit due to the attacker's increased difficulty in predicting future OSes relative to the diversity strategy.

As a result of these simulated experiments, it is recommended that when interaction with an attacker can be kept to a short duration, a diversity strategy is preferred. The specific length of the duration that is optimal depends upon the expected time it would take the attacker to develop the exploits.

4 Conclusion

We have developed a model of adaptive attacker strategy evolution and used it to investigate the strategies an attacker develops to overcome two temporal platform migration moving target defense strategies. The attacker–defender interaction has been modeled as a game in which a nonadaptive defender deploys a randomization or a diversity moving target defense. Against these dynamic platform scheduling policies, a population of attackers develop strategies specifying the temporal ordering of resource investments that bring zero-day exploits into existence to compromise the defender's system.

The results of this study have strong implications for real-world defenders. First, defenders deploying dynamic platform defenses and anticipating attacks over *difficult channels* (i.e., requiring persistence to succeed) should be particularly vigilant regarding the systems in their rotation-set with outlier status in attributes relevant to an attacker's success. It is these outlier systems that advanced attackers will devote the largest proportion of resources to compromising. Furthermore, our results suggest that diversity-maximizing defenses are most effective for short duration attacker/defender encounters. The crucial parameter in this regard is the time required for an attacker to bring exploits into existence versus the duration of the attacker's encounter with the defender's system.

Future directions of interest for investigation include the incorporation of noise into the attacker's observation model in order to bring the game scenario nearer to conditions likely to prevail for real-world attackers and the incorporation of an adaptive defender into our cyber game scenario.

References

Bastian, M., Heymann, S., & Jacomy M. (2009). *Gephi: An open source software for exploring and manipulating networks*. International AAAI conference on weblogs and social media, pp. 361–362.

Carter, K. M., Okhravi, H., & Riordan, J. (2014). Quantitative analysis of active cyber defenses based on temporal platform diversity. *arXiv preprint arXiv:1401.8255.*

Colbaugh, R., & Glass, K. (2012). *Predictability-oriented defense against adaptive adversaries.* Proceedings of the 2012 IEEE international conference on systems, man, and cybernetics, pp. 2721–2727.

Holland, J. H. (1975). *Adaptation in natural and artificial systems*. Ann Arbor: The University of Michigan Press.

MATLAB R2013b. (2013). The MathWorks, Inc., Natick: Massachusetts.

Michalewicz, Z. (1996). *Genetic algorithms + data structures = evolution programs, 3E*. Berlin: Springer.

Miller, J. H. (1996). The coevolution of automata in the repeated Prisoner's dilemma. *Journal of Economic Behavior and Organization, 29*, 87–112.

Mitchell, M. (1996). *An introduction to genetic algorithms*. Cambridge: MIT Press.

Okhravi, H., Rabe, M., Mayberry, T., Hobson, T., Bigelow, D., Leonard, W., & Streilein, W. (2013). *Survey of cyber moving target techniques*. MIT Lincoln Laboratory Technical Report, 1166.

Okhravi, H., Hobson, T., Bigelow, D., & Streilein, W. (2014). Finding focus in the blur of moving-target techniques. *IEEE Security & Privacy, 12*(2), 16–26.

Schleimer, S., Wilkerson, D.S., & Aiken, A. (2003). *Winnowing: Local algorithms for document fingerprinting*. Proceedings of the 2003 ACM SIGMOD international conference on management of data, pp. 76–85.

Tambe, M. (2012). *Security and game theory: Algorithms, deployed systems, lessons learned.* New York: Cambridge University Press.

Wilensky, U. (1999). NetLogo, Center for connected learning and computer-based modeling. http://ccl.northwestern.edu/netlogo/, Evanston: Northwestern University.

Winterrose, M.L., & Carter, K. M. (2014). *Strategic evolution of adversaries against temporal platform diversity active cyber defenses*. Proceedings of the 2014 Symposium on Agent Directed Simulation. Society for Computer Simulation International, 2014.

Deep Reinforcement Learning for Adaptive Cyber Defense and Attacker's Pattern Identification

Ahmad Hoirul Basori and Sharaf Jameel Malebary

1 Introduction

In 2018, the Identity Theft Resource Centre in the USA has recorded around 560 million data with more than 1000 attacks. The highest percentage is in the business sector that reaches 46% breach out of 94.7% record of data followed by medical/ healthcare with 29.3% breach (ITRC 2018; PwC 2017). It is widely known in the industry that hackers are capable to do exploitation in a very short period and cause great damage. The cost to fixed the system into normal state will require great amount of effort and time. It may take up to 146 days to rescue the system and fix the vulnerabilities (Zeichick et al. 2017). The cost of this repair is not cheap and can reach approximately $4 million. Additionally, such attacks will cause a huge loss to the company and they even become more concentrated with this attack behavior. This condition will drag the company in the dilemma situation where the attackers can launch their defense and target the defenseless companies. Therefore, it needs an innovative solution that can predict or identify the pattern of the attacker's behavior in order to launch a counterattack. The structure of the paper is started with introduction toward cybersecurity followed by related works from the previous researcher in the second section. Next, Sect. 3 discusses the material and research method, while Sect. 4 presents the result and discussion. Finally, Sect. 5 provides a conclusion and future works.

A. H. Basori (✉) · S. J. Malebary
Faculty of Computing and Information Technology Rabigh, King Abdulaziz University, Rabigh, Makkah, Kingdom of Saudi Arabia
e-mail: abasori@kau.edu.sa; smalebary@kau.edu.sa

S. K. Shandilya et al. (eds.), *Advances in Cyber Security Analytics and Decision Systems*, EAI/Springer Innovations in Communication and Computing,
https://doi.org/10.1007/978-3-030-19353-9_2

2 Related Works

The latest report by Cisco states that half of the attacks that occurred in 2017 has caused great loss and costs up to $500,000. Therefore, a company should give extra attention to this kind of cyberattacks. The IoT has dragged hacker attention to attack and penetrate the company data that usually are saved in the store. The other study focused on implementing AI for securing the operation that happens between humans and machines. They focused on data mining and machine learning approach to identify the intrusion detection pattern or behavior (Wilkins 2018; Cisco 2018). Furthermore, the researchers also tried to use a learning-based technique for spam filtering and malware detection (Buczak and Guven 2015). The growth of machine learning has touched a wide area and even capable to be used as automatic detection and analysis that can support the security analysts. Some of machine learning techniques are presented to see their fitness toward intrusion detection, malware counterattack and detection, and even spam filtering (Blanzieri and Bryl 2008; Gardiner and Nagaraja 2016).

As we know intrusion detection objective is to realize forbidden actions inside a computer or network via intrusion detection systems (IDS). IDS inside network has been implemented in most of the modern networks. Their working methods are based on the pattern of the attack history/log (Apruzzese et al. 2018; Pierazzi et al. 2017). The latest IDS can detect an anomaly and potential threat or even do classification with machine learning approaches. There are a lot of detection problems such as botnets detection and domain generation algorithms (DGA). The botnet is known as the network that is fully controlled by hackers and can be used for illegal actions, e.g., attacking/infecting another network. DGA is automatic domain production that is usually used by the infested server to produce communication with a server outside their network. Reinforcement learning is also capable to be used as spam and phishing detection by adding learning process toward email types, hidden malware, or even compromised links. Walta et al. (2018) give a statement that cybersecurity on social media is easily widespread and requires more attention from authorizing person/organization.

They are proposing an intelligent solution that considers the personality of a person in social media. The system will use profile images as an attribute for the machine learning process. The learning process will enable a system to do gender classification as well as potential deception detection with more precision (Walta et al. 2018). The deception in psychopathological point of view is referring to changing the name or profile picture in the social media due to dissatisfaction toward themselves which may lead to deception (Stanton et al. 2016). Hancock (2007) categorized the deception in social media which can be through identity or message. He also argued that deception on social media is tougher to be detected compared to real-world deception.

Most of the users usually don't pay enough attention toward the complete URL or a valid URL of a certain website. They simply click the URL which is posted in social media. This action triggers a phisher to steal personal information that may contain a username, password, or both (Gupta et al. 2018). The other researcher has come up with a proposed solution with a real-time anti-phishing system that used seven classification algorithms and NLP (neuro-linguistic programming) that is capable of detecting new website and is also compatible with any language and third-party systems (Sahingoza et al. 2019).

This paper presents an innovative framework that involved the reinforcement learning algorithm to analyze and identify the learning pattern of attackers. The proposed framework acts as an intrusion detection system, malware analysis, and spam or phishing detection approaches that can prevent attackers from manipulating the system further.

3 Research Method and Material

The reinforcement learning algorithm is widely known as its capability on providing an agent to learn and adapt toward a certain goal. The algorithm focused on achieving the goal over many stages. Reinforcement learning is capable of solving the complex problem by delaying the result of a process toward immediate actions (SkyMind AI 2019). The reinforcement learning will gain optimum result when it is implemented in the real-world problem, where the random/illogical number of action will emerge (Sutton and Barto 2017).

The reinforcement learning working scheme can be depicted from the concepts of agents with their states, environment, rewards, and action as shown in Fig. 1.

The agent playing a role as a subject during interaction can be a device, human, or even a character in a video game. The action (A) is a collection of a feasible movement that they can perform during the interaction, while the environment is a situation or a place where agents move. A state (S) is the condition where an agent positioned themselves and, lastly, the reward (R) is a response that represents the success or failure rate of an agent (SkyMind 2019; Sutton and Barto 2017). The main idea proposed in this paper is focused on how we provide a generic framework that is able to manage the learning process of a "guard" and can block and detect some attacks. This framework is adapted from the Dyna model (Dyna: Integrating

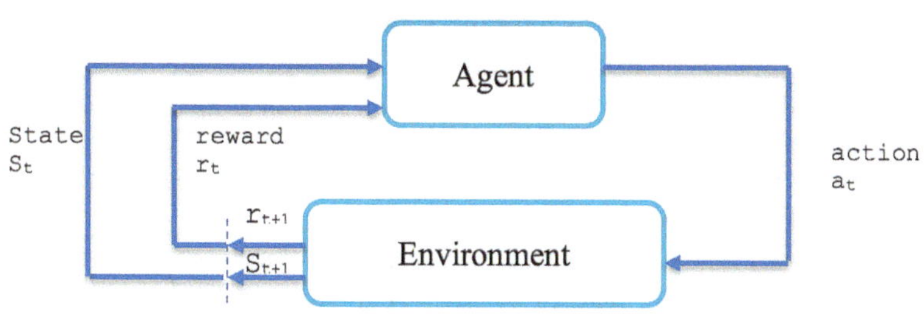

Fig. 1 A typical framework of reinforcement learning (SkyMind AI 2019; Sutton and Barto 2017)

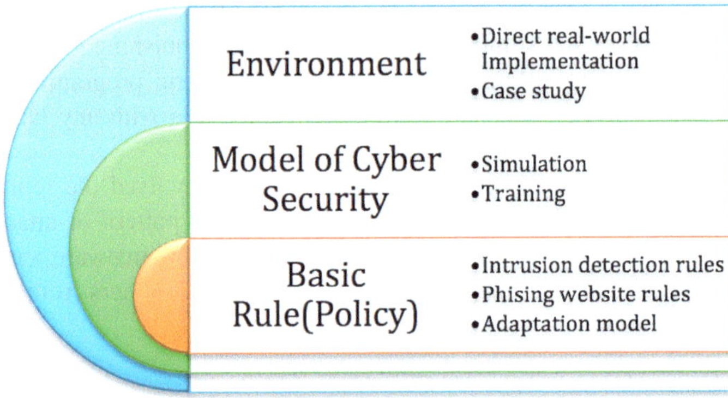

Fig. 2 A generic framework of cybersecurity with reinforcement learning adopted from Dyna model (Sutton and Barto 2017)

Planning, Acting, and Learning) of Suton and Barto (2017). So basically, we are designing the three major components such as environment, model, and policy/rules.

Figure 2 shows that policy/basic rules for security/intrusion detection are the basis of how the guard agent will act or respond toward an attack. It also consists of adaptation model that will enhance the "guard" agent to adapt toward certain intrusion situation. The second component will focus on the model of the security itself, in reinforcement learning; it will be dedicated for model learning and planning revision through simulation or training. The objective equation of reinforcement learning can be defined as Eq. 1:

$$\sum_{t=0}^{t=\infty} \gamma^t r\left(x(t), a(t)\right) \tag{1}$$

Finally, the environment will lead the "guard" agent to perform direct reinforcement learning experience in the real-world implementation or a case study simulation. This proposed framework provides new approaches and benefits for the adaptive cyber defense and attacker identification. The result of this framework is an idea that can be used as a reference for intrusion detection, malware, or even spam-phishing research analysis using reinforcement learning.

As the initial process of our proposed method, we tested the dataset for website phishing collected by Rami, McCluskey, and Thabtah (2014). The dataset contains several features such as IP address, URL redirect, the age of domain, website traffic, etc. The loaded data in the system is shown in Fig. 3.

Figure 3 illustrates the loaded dataset that consists of more than eleven thousand records and more than 30 features of data. This data will be evaluated based on several classifiers as a comparator with the proposed framework.

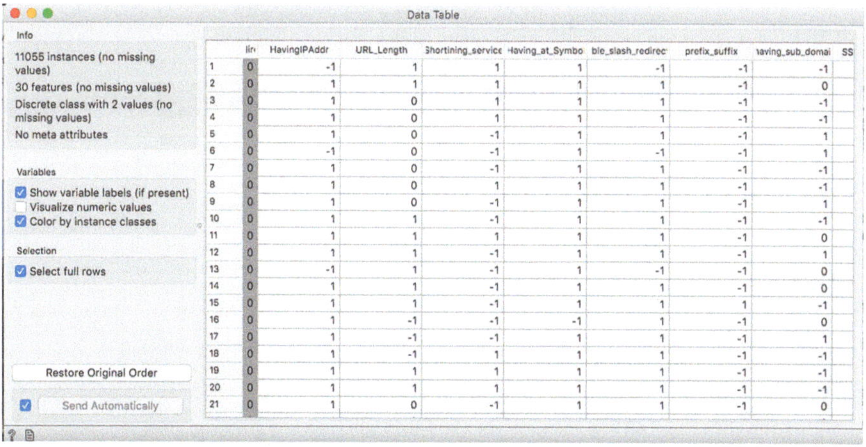

	lin	HavingIPAddr	URL_Length	Shortining_service	Having_at_Symbo	ble_slash_redirec	prefix_suffix	having_sub_domai	SS
1	0	-1	1	1	1	-1	-1	-1	
2	0	1	1	1	1	1	-1	0	
3	0	1	0	1	1	1	-1	-1	
4	0	1	0	1	1	1	-1	-1	
5	0	1	0	-1	1	1	-1	1	
6	0	-1	0	-1	1	-1	-1	1	
7	0	1	0	-1	1	1	-1	-1	
8	0	1	0	1	1	1	-1	-1	
9	0	1	0	-1	1	1	-1	1	
10	0	1	1	-1	1	1	-1	-1	
11	0	1	1	1	1	1	-1	0	
12	0	1	1	-1	1	1	-1	1	
13	0	-1	1	-1	1	-1	-1	0	
14	0	1	1	-1	1	1	-1	0	
15	0	1	1	-1	1	1	1	-1	
16	0	1	-1	-1	-1	1	-1	0	
17	0	1	-1	-1	1	1	-1	1	
18	0	1	-1	1	1	1	-1	-1	
19	0	1	1	1	1	1	-1	-1	
20	0	1	1	1	1	1	-1	-1	
21	0	1	0	-1	1	1	-1	0	

Fig. 3 Loaded dataset for web phishing dataset

4 Result and Discussion

As mentioned before the dataset is being tested toward different classifiers to see the generic behavior of dataset. So, based on the proposed framework given in Fig. 3, the initial stage is started by giving the basic rules for phishing, e.g., the IP address Suton and Barto (2017):

$$\text{Rule : IF} \begin{cases} \text{If The Domain Part has an IP Address} \rightarrow \text{Phishing} \\ \qquad\qquad\text{Otherwise} \rightarrow \text{Legitimate} \end{cases}$$

Afterward, the given model is trained and simulated through several scenarios to observe the fitness of the model. Currently, we didn't reach the real-life testing yet due to short of time. However, the simulation has been conducted by comparison with two different classifiers and found that reinforcement learning is high correlation when it is matched with neural network and also superior compared to other classifiers such as CN2 induction and SVM (support vector machine) algorithms as shown in Fig. 4.

Figure 4 shows that the proposed reinforcement learning framework is fit with a neural network with confidence rate around 95% then followed by CN2 rule and SVM with 76% confidence rate.

Figure 5 represents the neural network has shown a promising percentage with correct prediction around 96% even though the CN2 rule induction is a little bit better in this case, while SVM got very low of predicted values that can lead to unfitness of SVM for this kind of dataset. The evaluation is then continued by ROC (receiver operating characteristic) curve which displays the stability of the neural network approach compared to SVM that has linear escalation since the beginning of the curve, while CN2 rule induction only has linear increment for a while and then remains stable as neural network, as shown in Fig. 6.

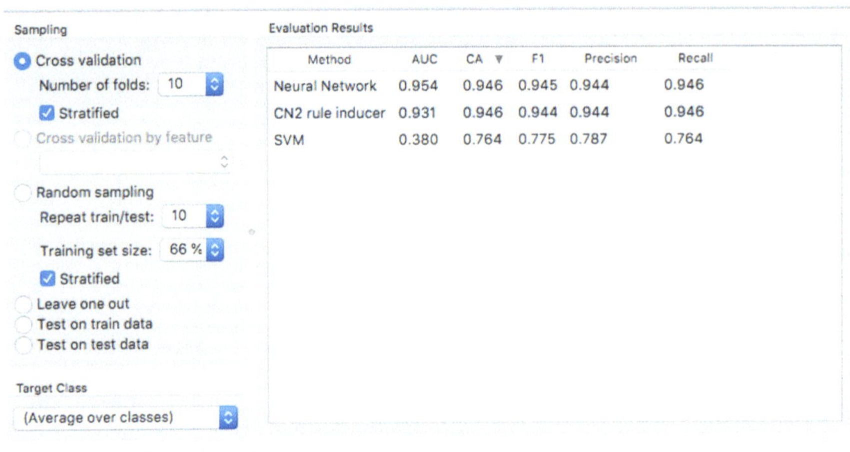

Fig. 4 Training result

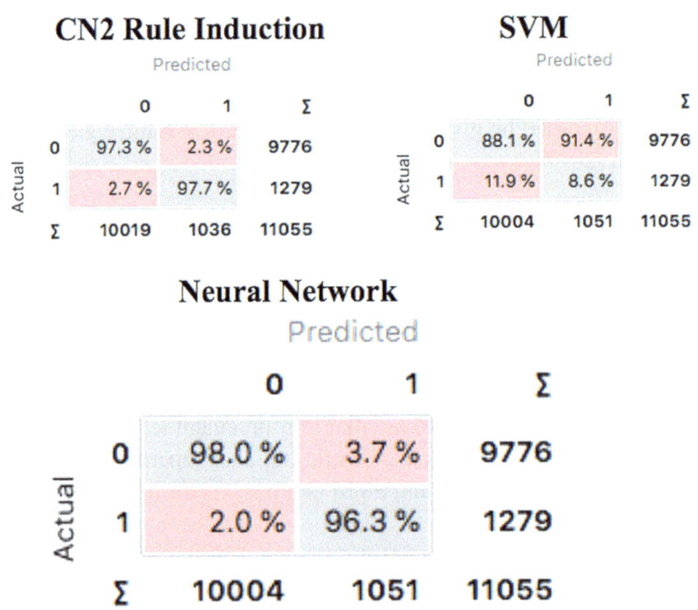

Fig. 5 Confusion matrix for the data set evaluation

Figure 7 is another representation for the dataset evaluation toward reinforcement learning framework; the neural network remains consistent since the beginning of the graph, while CN2 induction is still lower compared to the neural network and SVM is the lowest among them.

Figure 8 is another graph evaluation based on a calibration plot. It is shown that the neural network has a similar trend with CN2 while SVM is more unstructured in this matter. Neural network illustrates the higher predicted probability may be in

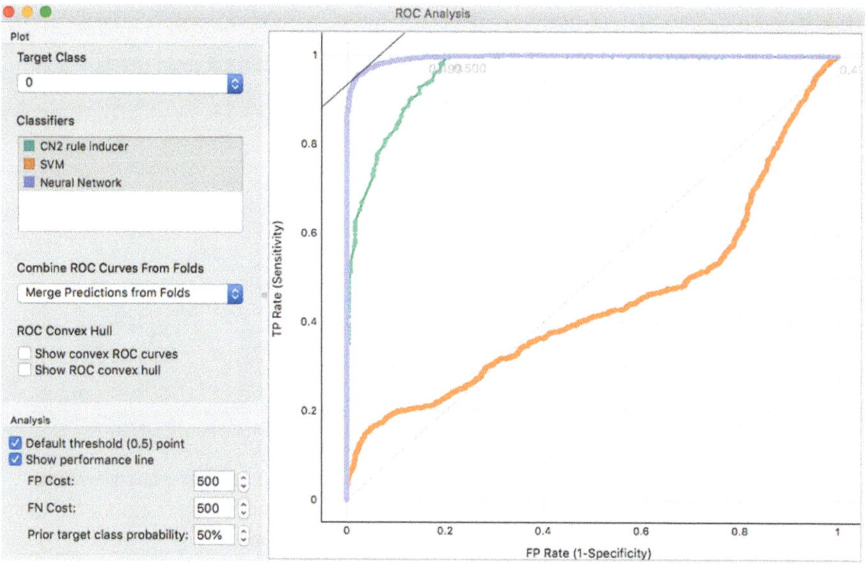

Fig. 6 ROC analysis

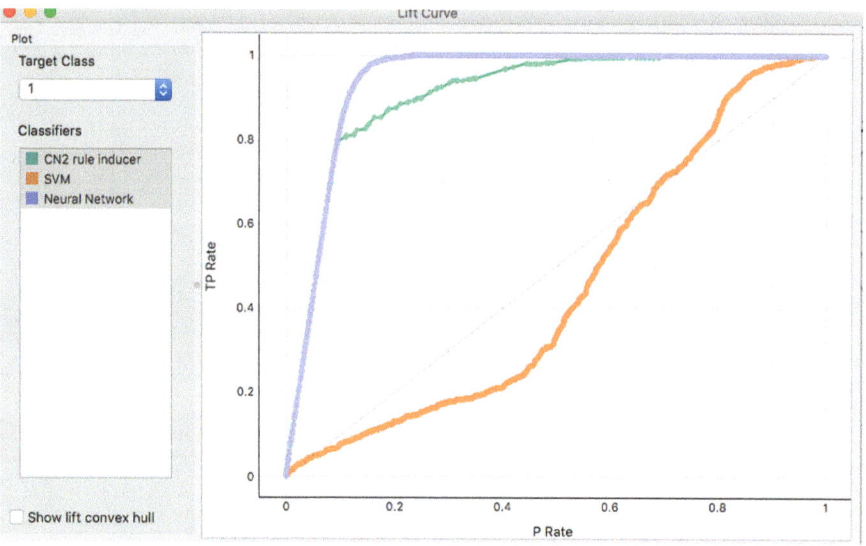

Fig. 7 Lift curve analysis

line along with observation average value. SVM initiates the observation when it reaches predicted probability more than 0.2 with observation average 0.7; then it raised a little bit until 0.9, then remained stable, or slightly declined.

Figure 9 argued that the dataset with non-attack has overcome the attack features with higher density value. This means the proposed framework has successfully trained and simulated to overcome and identify the attacker's pattern that produces a similar sequence of attack. In addition, Fig. 10 is designed to describe the general

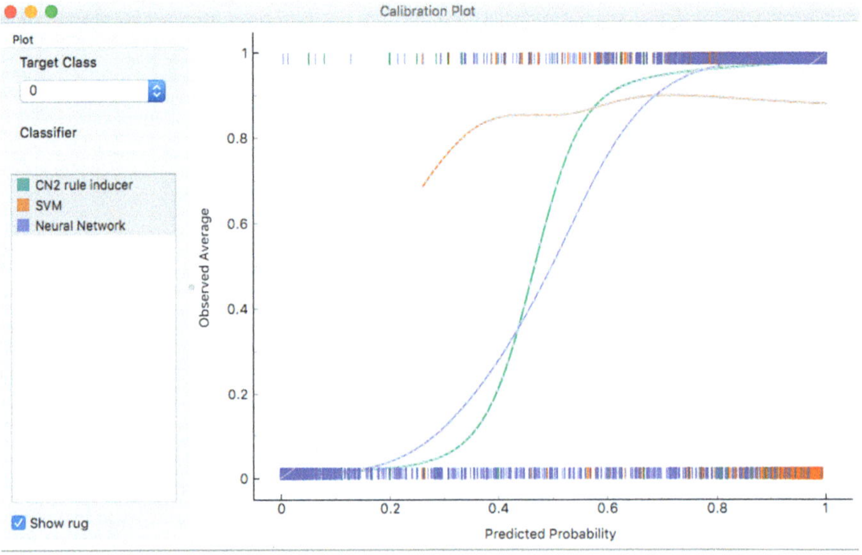

Fig. 8 Calibration plot analysis

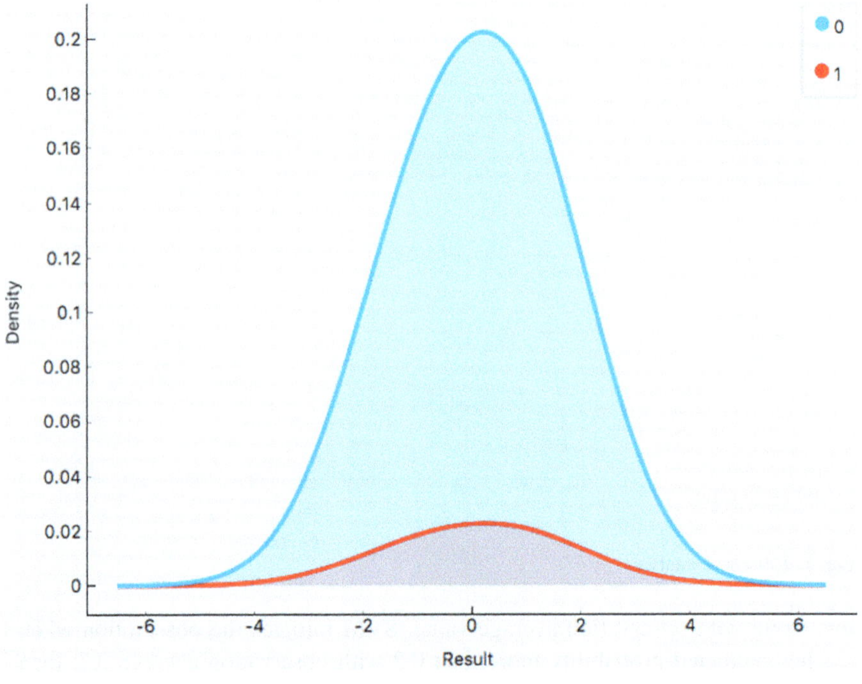

Fig. 9 Density diagram of the result features of dataset

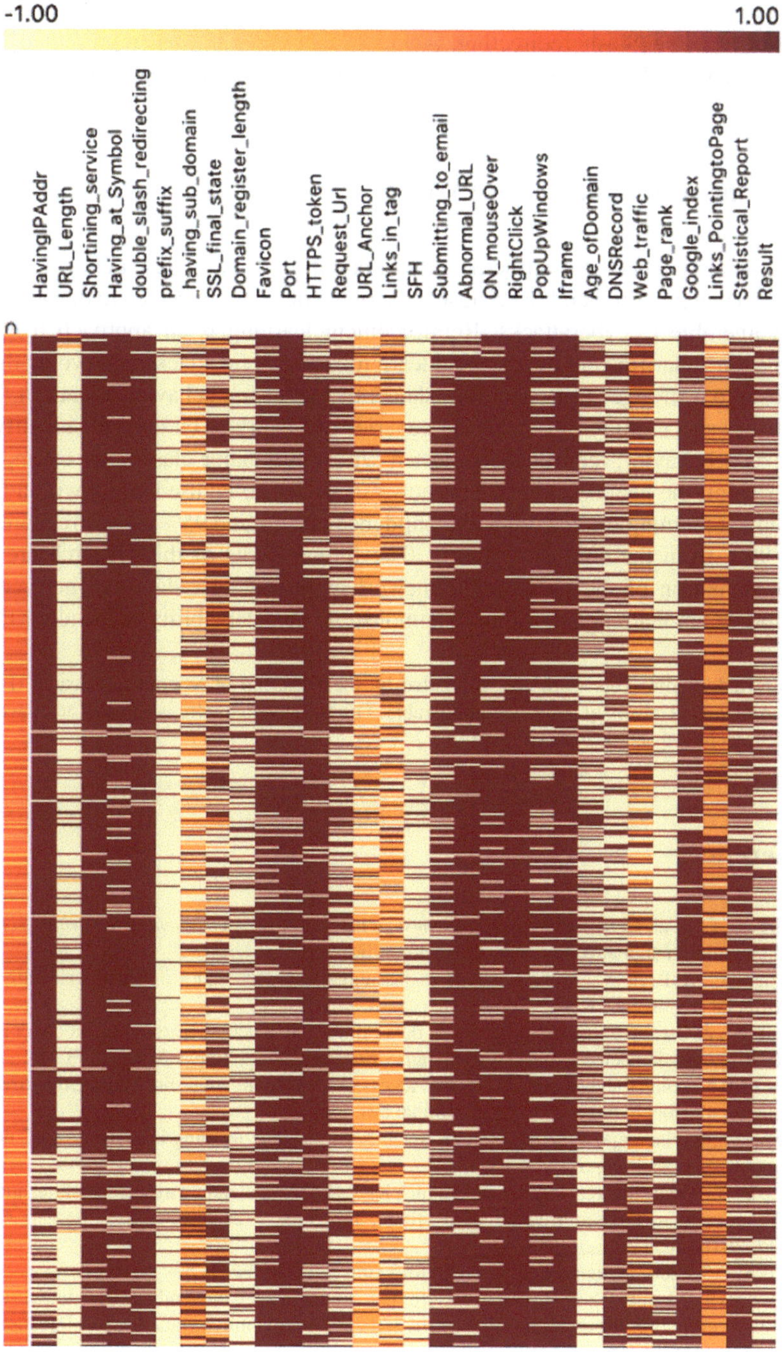

Fig. 10 Heat map analysis

result from the experiment. As shown in Fig. 10, variables URL_length, Having_at_symbol, Double_Slash_redirecting, Abnormal_URL have been represented by a dark red colour. This means it mostly contributes toward phishing website attack.

5 Conclusion and Future Works

Cybersecurity indeed has attracted a lot of attention from the researcher, academician, or even practitioner. It's because most of the devices connected to the networks are vulnerable to cyberattacks. Reinforcement learning is an approach that can enable the "guard" agent to adapt and learn toward certain attacks and evolve themselves into a stronger entity that may detect the attacker's behavior or pattern. This paper has successfully presented the generic framework to be used for identifying attacker's behavior; then it is also tested with three major classifiers. During the experiment, the neural network has shown its fitness to be used along with a reinforcement learning framework with confidence rate around 95% and correct prediction 96% from the confusion matrix. The future works of this research are full implementation of the proposed framework with real-case data and hard-coded reinforcement learning algorithm to demonstrate the adaptability of the algorithm.

Acknowledgments This work was supported by the Deanship of Scientific Research (DSR), King Abdulaziz University, Jeddah, Saudi Arabia. The authors, therefore, gratefully acknowledge the DSR technical and financial support.

References

Apruzzese, G. et al. (2018). *On the effectiveness of machine and deep learning for cyber security*, the 10th international conference on cyber conflic.

Blanzieri, E., & Bryl, A. (2008) A survey of learning-based techniques of email spam filtering, Artificial Intelligence Review.

Buczak, A., & Guven, E. (2015). A survey of data mining and machine learning methods for cyber security intrusion detection. *IEEE Communications Surveys & Tutorials, 18*, 1153.

Cisco. (2018). Annual cybersecurity report, Accessed May 2018. www.cisco.com/c/dam/m/digital/elq-cmcglobal/witb/acr2018/acr2018final.Pdf

Gardiner, J., & Nagaraja, S. (2016). On the security of machine learning in malware C8C detection. *ACM Computing Surveys, 49*(3), 59.

Gupta, B. B., Arachchilage, N. A. G., & Psannis, K. E. (2018). Defending against phishing attacks: Taxonomy of methods, current issues and future directions. *Telecommunication Systems, 67*(2), 247–267.

Hancock, J. T. (2007). *Digital deception*. Oxford University Press.

ITRC. (2018). Identity Theft Resource Centre data breach report, November 2018. https://www.idtheftcenter.org/wp-content/uploads/2018/12/2018-November-Data-Breach-Package.pdf Accessed Dec 2018.

Pierazzi, F., et al. (2017). Scalable architecture for online prioritization of cyber threats. In *International conference on cyber conflict (CyCon)*.

PwC. (2017). Information security breaches survey', Accessed Apr 2017. www.pwc.co.uk/services/audit-assurance/insights/2015-information- security-breaches-survey.html.

Rami, M., McCluskey, T. L., & Thabtah, F. A. (2014). Intelligent rule based phishing websites classification. *IET Information Security, 8*(3), 153–160. ISSN 1751-8709.

Sahingoza, O. K., Buberb, E., Demirb, O., & Diric, B. (2019). Machine learning based phishing detection from URLs. *Expert Systems with Applications, 117*(2019), 345–357, Elsevier.

SkyMind AI. (2019), https://skymind.ai/wiki/deep-reinforcement-learning, Last Accessed Jan 2019.

Stanton, K., Ellickson-Larew, S., & Watson, D. (2016). Development and validation of a measure of online deception and intimacy. *Personality and Individual Differences, 88*, 187–196.

Sutton, R. S., & Barto, A. G. (2017). *Reinforcement learning: An introduction, a Bradford book.* Cambridge/Massachusetts/London: The MIT Press.

Walta, E. V. D., Eloffa, J. H. P., & Grobler, J. (2018). Cyber-security: Identity deception detection on social media platforms. *Computers & Security, 78*(2018), 76–89, Elsevier.

Wilkins, J. (2018). Is artificial intelligence a help or hindrance? *Network Security, 2018*(5), 18–19. https://doi.org/10.1016/S1353-4858(18)30046-1.

Zeichick, A., Talwar, R., & Koury, A. (2017). AI in IT security, Network Security, https://www.imperva.com/blog/will-ai-change-the-role-of-cybersecurity/., Last Accessed Feb 2019.

Dynamic Recognition of Phishing URLs Using Deep Learning Techniques

S. Sountharrajan, M. Nivashini, Shishir K. Shandilya, E. Suganya, A. Bazila Banu, and M. Karthiga

1 Introduction

Phishing is like a spam in which the invader tries to learn confidential information such as login credentials or account details by sending a trustworthy entity through email or other transmission modes. Usually, a victim obtains a message which looks like it was sent by a recognized person/contact or association. The message contains malicious software which may crack confidential information present in the user's desktop. In addition, the malicious link attached with the body of the email message may use the organization's logo and genuine contents for spoofing.

Phishing techniques are considered to be the most serious threats to banking and financial sectors. Leading IT companies got duped out million dollars through an email phishing scheme when a hacker impersonated a computer-parts vendor. According to the Federal Bureau of Investigation, invaders made off with at least $676 million last year. Patrick Peterson, founder and CEO of email security-based company known as Agari located at San Mateo, California, said that email scam hits the largest count, because usage of email is essential nowadays. Phishers pretend to use social media for transmitting the malicious content and other open

S. Sountharrajan (✉) · E. Suganya
VIT Bhopal University, Bhopal, India

M. Nivashini
Anna University, Chennai, India

S. K. Shandilya
School of Computing Science & Engineering, Vellore Institute of Technology, VIT Bhopal University, Bhopal, Madhya Pradesh, India

A. Bazila Banu · M. Karthiga
Bannari Amman Institute of Technology, Coimbatore, India

S. K. Shandilya et al. (eds.), *Advances in Cyber Security Analytics and Decision Systems*, EAI/Springer Innovations in Communication and Computing,
https://doi.org/10.1007/978-3-030-19353-9_3

sources, containing social networks like LinkedIn, Facebook, and Twitter, to retrieve saved contents about the target's private and labor history and his actions.

Pre-phishing attack inspection can discover contact number, organization address, and electronic message addresses of possible targets, as well as material about their social group and the names of higher officials in their organizations. This evidence can then be used to draft an authentic email. Targeted attacks are conceded out by determined peril groups with a phishing email containing a spiteful link or attachment. Although many phishing electronic messages are poorly drafted and easily identified as fake email, people involved in cybercriminal activity use the same technique to recognize the messages. Victorious phishing messages, typically characterized as being from a famous concern, are complicated to differentiate from reliable messages because a phishing electronic message misinterprets the commercial logos, trusted tools, and symbols of the concern. Spiteful connections inside phishing messages are typically intended to influence it to seem like they go to the caricature association. The use of sub-domains and similar URLs is common guiles which make the Internet users trust the messages.

To summarize the phishing activity, users should avoid submitting their credentials to a non-trusted site. Each victorious pre-phishing recon attack will give an invader a profile to be used in future context-aware attacks. This profile contains the recipient's email address, name, permanent residential address, contact number, employer details, mother's maiden name, friend and business partners' details, saved password, history of the websites visited by the user, and more.

1.1 Problem Statement

The answer for distinguishing phishing is when experiencing low recognition exactness and high false caution rate especially when novel phishing procedures are taken care of by the trespassers. Also, the most well-known system utilized, the blacklist-based strategy, is awkward in finding hair-raising phishing assaults since enrolling new space has turned out to be less demanding and there is no far-reaching blacklist database accessible for illuminating the unapproved areas. Additionally, such sort of point-by-point data about the unapproved areas can't be distinguished in the beginning. What's more, page content assessment has been utilized to determine the false negative (FN) issues and so as to adjust the vulnerabilities of the smelly records. Be that as it may, page content examination calculations give distinctive ways to deal with identifying phishing sites with fluctuating degrees of exactness. Be that as it may, there exist a wide scope of issues in finding the answer for identifying phishing URL. The following list of queries needs to be addressed in order to evolve the more suitable technique:

1. How to practice unrefined dataset for phishing detection?
2. What are the methods to amplify discovery rate for phishing website algorithms?

3. What is the procedure to decrease false negative rate for phishing website algorithms?
4. How to determine the finest combination of classifiers in order to produce accurate recognition rate of phishing website?

1.2 Background and Motivation

First, lack of extensive sources of squatting domains is considered as the major root cause of this investigation. It is demanding to capture an inclusive list of squatting domains that are potentially impersonating rightful brands and online services. Introducing the background of phishing and techniques used for preprocessing the raw data is essential to find the root cause of the problem. The unstructured data about the URL (e.g., literary portrayal) need to be identified with proper extraction technique.

The role of preprocessing and feature extraction needs to be applied to extract the flowing features:

- having_IP_Address
- URL_Length
- Shortining_Service
- having_At_Symbol
- double_slash_redirecting
- Prefix_Suffix
- having_Sub_Domain
- SSLfinal_State
- Domain_registeration_length
- Favicon
- Port
- HTTPS_token
- Request_URL
- URL_of_Anchor
- Links_in_tags
- SFH
- Submitting_to_email
- Abnormal_URL
- Redirect
- on_mouseover
- RightClick
- PopUpWidnow
- Iframe
- age_of_domain
- DNSRecord
- web_traffic
- Page_Rank
- Google_Index

- Links_pointing_to_page
- Statistical_report

Soft computing methods are progressively more being utilized to specify the detection of phishing URLs. Albeit supervised learning accommodates a vastly improved precision, unsupervised learning accommodates a quick and reliable way to deal with concentrate data from the information.

1.3 Challenges in Detecting Phishing URLs

Phishing is a semantic assault that uses electronic mediums to convey the content with natural languages like English, French, Chinese, Arabic, etc., to influence victims to perform convinced activities.

The challenges here are as follows:

- Computers have extraordinary complexity in decisively understanding the semantics of common dialects.
- Authentication and security instruments are as yet relying upon client-based inclination settings both from the web and program side.
- Phishing prevention methods are based on known context.
- Lack of Internet-based knowledge among the users accessing social media.

1.4 Objective of Study

Phishing URL uses are expanded by sidestepping frameworks. Shirking strategies are made by making lightweight properties that are dropped by winning characteristic choice procedures. In our methodology, phishing URL avoidance procedures are settled by dynamically choosing the attributes dependent on the kind of landing URLs as an alternative of utilizing just explicit set of chosen attributes by existing attribute selection methods. Therefore, in this paper, we emphasize on the determination of attributes, removed from the URLs utilizing DBM and SAE to find phishing assaults proficiently with low false positive rate and encounter ongoing demand with high identification rate utilizing DNN.

1.5 Scope and Limitation

The trouble in the proposed technique is preparing the whole DBM demonstration which is tedious with undirected associations between the layers. In the future, the previously mentioned inadequacy of the DBM model can be defeated utilizing deep belief network (DBN) which creates utilization of coordinated associations of the hubs between the layers. More deep learning calculations like recurrent neural network, convolutional neural network, and so on can be utilized in the phishing URL recognition. Additionally, vast scale ongoing dataset can be considered in building up the model. Arrangement by means of phishing URL identification as an administration or as a program augmentation should be possible to order any URL interface on the web.

2 Literature Review

2.1 Phishing URL Detection Methods

Phishing is a social building assault that goes for misusing the shortcoming found in framework forms as brought about by framework clients. For instance, a framework can be in fact sufficiently secured against secret word burglary; anyway, uninformed end clients may release their passwords if an aggressor requested that they refresh their passwords by means of a Hypertext Transfer Protocol (HTTP) interface, which at last compromises the general safety of the framework.

Since phishing assaults go for abusing shortcomings found in people (e.g., framework end clients), it is not easy to moderate them. For instance, like in (Sheng et al. 2010), end clients neglected to identify phishing assaults by 29% notwithstanding when prepared with the top performing client mindfulness program. Then again, programming phishing identification strategies are assessed against mass phishing assaults, which make their execution for all intents and purposes obscure with respect to focused types of phishing assaults. These constraints in phishing moderation strategies have for all intents and purposes brought about security breaks against a few associations including driving data security suppliers (Krebs 2011; Schneier 2011).

Because of the wide idea of the phishing issue, phishing identification overview starts by:

- Defining the phishing issue. Note that the phishing definition in the writing isn't steady, and hence, an examination of various definitions is introduced.
- Sorting hostile to phishing arrangements from the perspective of phishing exertion life cycle. This exhibits the different enemy of phishing arrangement classes, for example, discovery. It is essential to see the general enemy of phishing picture from abnormal state point of view before plunging into a specific procedure, specifically, phishing discovery methods (which is the extent of this overview).
- Displaying assessment measurements that are usually utilized in the phishing area to assess the execution of phishing discovery strategies. This encourages the examination between the different phishing identification methods.
- Displaying a writing study of hostile to phishing discovery strategies, which joins programming identification methods just as client mindfulness procedures that upgrade the location procedure of phishing assaults.

- Displaying an examination of the different proposed phishing identification strategies in the literature.

Definition

The definition of phishing assaults isn't reliable in the writing, which is because of the way the phishing issue widens and fuses fluctuating situations. For instance, as per PhishTank:

> Phishing is a fraudulent attempt, usually made through email, to steal your personal information.

The definition of PhishTank remains constant in various situations which, generally, spread most of phishing assaults.

Another definition is given by Colin Whittaker et al.:

> We define a phishing page as any web page that, without permission, alleges to act on behalf of a third party with the intention of confusing viewers into performing an action with which the viewer would only trust a true agent of the third party.

Colin Whittaker et al.'s (2010) definition plans to be more extensive than PhishTank's, and one might say that aggressors' objectives are never again limited to taking individual data from unfortunate casualties. Then again, the definition still limits phishing assaults to ones that follow up the interest of outsiders, which isn't in every case genuine.

History

As indicated by APWG (the Anti-Phishing Working Group), the word phishing was begat in 1996 because of social designing assaults against America Online (AOL) accounts through online con artists. The term phishing originates from the job fishing where one might say that fishers (e.g., aggressors) utilize a trap (e.g., socially designed messages) to fish (e.g., take individual data of unfortunate casualties). The "ph" substitution of the character "f" in fishing is because of the way that one of the most punctual types of hacking was against phone systems, named as phone phreaking. Accordingly, ph turned into a typical hacking character substitution of "f." As indicated by APWG, pilfered records through phishing assaults were likewise utilized as a cash amid programmers by 1997 to exchange hacking programming in return of the pilfered records. Phishing assaults were truly begun by taking AOL accounts and throughout the years passed into assaulting progressively gainful targets, for example, online banking and Internet business administrations. As of now, phishing assaults target not just framework end users but also specialized workers at specialist co-ops and may convey modern procedures, for example, MITB (Man-in-the-Browser) assaults.

Phishing Motives

As indicated by Weider D. et al. (2008), essential intentions on phishing assaults, since an assailant's viewpoint, are:

- Monetary profit: phishers can utilize pilfered financial certifications to their financial benefits.

- Character stowing away: rather than utilizing stolen personalities specifically, phishers may pitch the personalities to others whom may be culprits looking for approaches to shroud their personalities and exercises (e.g., buy of merchandise).
- Distinction and reputation: phishers may assault unfortunate casualties for companion acknowledgment.

Difficulties

Since the phishing issue exploits human numbness or naivety concerning their cooperation with electronic correspondence channels (e.g., email, HTTP, and so forth), it's anything but a simple issue for all time tackle. The majority of the proposed arrangements endeavor to limit the effect of phishing assaults. From an abnormal state point of view, there are two commonly regularly proposed answers to alleviate phishing assaults:

Client instruction: the human is taught trying to upgrade their classification exactness to effectively recognize phishing posts and afterward do appropriate activities on the accurately classified phishing messages, for example, revealing assaults to framework executives.

Programming upgrade: the product is improved to more readily group phishing messages for the human or give data in an increasingly evident manner so the human would have less opportunity to disregard it.

Alleviation of Phishing Attacks: An Overview

Because of the wide idea of the phishing issue, we find it critical to picture the life cycle of the phishing assaults, and depending on that, order hostile to phishing arrangements. When the phishing assault is distinguished, various activities could be connected against the battle. As per survey of the writing, the accompanying classes of methodologies exist:

A. *Location Approaches*

Any enemy of phishing arrangement expects to recognize or group phishing assaults as identification arrangements. This incorporates:

Client preparing approaches – end clients can be taught to more readily comprehend the idea of phishing assaults, which at last points them into effectively distinguishing phishing and non-phishing messages. Be that as it may, client preparing approaches go for upgrading the capacity of end clients to identify phishing assaults, and along these lines, we sort them under "identification."

Programming classification approaches – these moderation approaches go for characterizing phishing and authentic messages for the benefit of the client trying to overcome any issues that are left because of the human mistake or numbness. This is a vital hole to connect as client preparing is costlier than robotized programming classifiers, and user training may not be doable in certain situations (e.g., when the client base is immense, e.g., PayPal, eBay, and so forth.).

The execution of recognition methodologies can be improved amid the education period of a classifier (regardless of the classifier, a human or programming). Location procedures not just help in specifically shielding end clients from falling unfortunate casualties to phishing efforts, however can likewise help in improving phishing honeypots to confine phishing junk from non-phishing junk.

B. *Hostile Defense Approaches*

Hostile protection arrangements intend to render phishing efforts pointless for aggressors by disturbing the phishing efforts. This is frequently accomplished by spreading phishing sites with phony qualifications so the aggressor would have a tough time to identify the genuine certifications.

Monitored system baits that can be utilized to occupy assailants from assaulting significant assets, give early admonitions about new assault inclines, or empower inside and outside examination of the performed assaults (Fig. 1).

Two best examples are:

BogusBiter (Yue and Wang 2008) – the program toolbar that succumbs counterfeit data in HTML frames at whatever point a phishing site is experienced. As indicated by BogusBiter, the identification of phishing sites is finished by different instruments. As such, rather than essentially appearing cautioning message to the end client at whatever point a phishing site is visited, BogusBiter additionally succumbs counterfeit information into HTML types of the visited phishing site. Succumbing counterfeit information into the HTML shapes is expected to disturb the relating phishing efforts, with expectation that such phony information may lead the assailants' assignment of identifying right information (amid the phony information) more tough. This is an endeavor to spare the pilfered

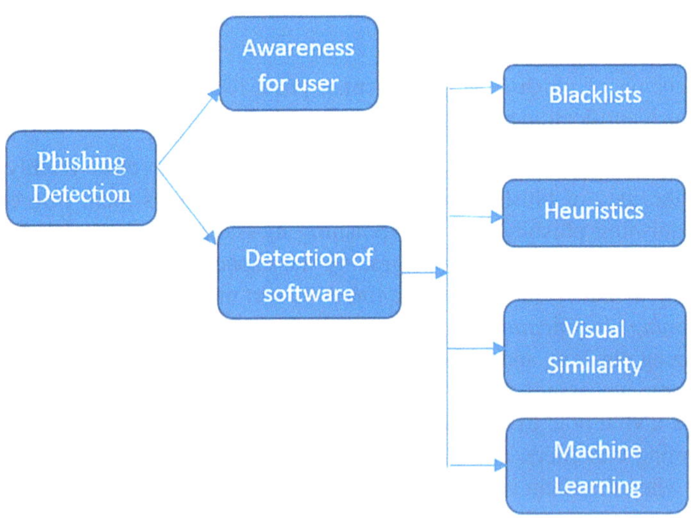

Fig. 1 Phishing detection approaches

accreditations of different clients that have been caught by the phishing effort by debasing the caught outcomes with counterfeit information. Be that as it may, the impediments are:

- Toolbars should be introduced on a sufficiently wide client base to render this successful.
- On the off chance that the client base is sufficiently wide, BogusBiter may lead to denial-of-service (DOS) attack against servers that have authentic mutual facilitated sites also, essentially on the grounds that one of the mutual web hosts has a phishing content.
- Expanded transmission capacity request.
- Non-standard HTML frames are not identified by BogusBiter.
- The exact viability of such kind of arrangement isn't precisely estimated.

Humboldt (Knickerbocker et al. 2009) – similar to BogusBiter; then again, actually, BogusBiter depends on entries from end client customers, while Humboldt depends on disseminated and devoted customers in the Internet rather than end user's toolbars visiting phishing locales, notwithstanding a system to abstain from triggering DOS floods against servers. This makes Humboldt progressively successful against phishing sites because of the extra incessant accommodation of information to phishing pages. Restrictions include:

- Expanded data transmission request.
- Humboldt does not recognize non-standard HTML frames.
- The experimental adequacy of such arrangement isn't precisely estimated.

Albeit hostile barrier methodologies can hypothetically make the aggressors' undertaking more difficult in finding an unfortunate casualty's close to home data, it isn't known how difficult it truly progresses toward becoming.

C. *Revision Approaches*

When a phishing effort is recognized, the redress procedure can start. On account of phishing assaults, adjustment is demonstration of bringing phishing assets down. This regularly accomplishes by detailing assaults to the service-providing agencies. Phishing efforts regularly depend on assets, for example:

- Sites – it would be a common web possessed by phisher, an authentic site with phishing materials transferred to it, or various contaminated end client workstations in a botnet (a botnet is various tainted PCs constrained by assailants for malevolent use).
- Email messages – it would be sent from an assortment of sources, for example, email service provider (ESP) (Gmail, Rediffmail, Yahoo mail, Hotmail), Simple Mail Transfer Protocol (SMTP) transfers, or tainted end user machines which are a piece of a botnet.
- Long-range informal communication administrations – web 2.0 administrations (e.g., Facebook, Twitter) could be utilized to convey publically designed posts to influence unfortunate casualties to uncover their PINs.

- Public Switched Telephone Network (PSTN) and Voice over IP (VoIP) – like different types of phishing assaults, assailants endeavor to induce exploited people to perform activities. Nonetheless, the thing that matters is that assailants endeavor to misuse spoken discoursed so as to gather information (instead of tapping on connections).

So as to address such conduct, mindful gatherings (e.g., service suppliers) endeavor to bring the assets low, for instance:

- Expulsion of phishing content from the websites or suspension of facilitating administrations
- Interruption of email accounts, Simple Mail Transfer Protocol transfers, and Voice over IP administrations
- Follow back and stoppage of blacklists

D. *Counteractive Action Approaches*

The "counteractive action" of phishing assaults can confound, as it can point diverse things relying upon its specific situation:

- Counteractive action of clients from falling injured individual – for this situation, phishing identification methods will likewise be viewed as anticipation strategies. In any case, this isn't the setting we allude to when "counteractive action" is referenced in this overview.
- Counteractive action of assailants from the beginning of phishing efforts – for this situation, claims and punishments against aggressors by law enforcement agencies (LEAs) are taken into consideration as avoidance methods.

Assessment Metrics

In parallel classification issue, the objective is to recognize phishing occurrences in a data collection set with a blend of phishing and authentic cases, among which four classification conceivable outcomes exist. Look at the perplexity framework exhibited in Table 1 for subtleties, where N is quantity of phishing examples which are accurately classified as phishing, $N_{P->P}$ is quantity of genuine cases which are erroneously classified as phishing, $N_{L->P}$ is quantity of phishing occasions which are inaccurately classified as real, $N_{P->L}$ is quantity of real occurrences which are mistakenly classified as real, and $N_{L->L}$ is quantity of authentic cases which are effectively classified as real.

The equations below show the commonly used phishing evaluation metrics.

True positive (TP) rate – processes the amount of accurately identified phishing assaults in connection to all current phishing assaults as exhibited in Eq. 1.

Table 1 Confusion matrix for classifying phishing

	Phishing	Legitimate
Is phishing?	$N_{P->P}$	$N_{P->L}$
Is legitimate?	$N_{L->P}$	$N_{L->L}$

$$\mathrm{TP} = \frac{N_{\mathrm{P->P}}}{N_{\mathrm{P->P}} + N_{\mathrm{P->L}}} \tag{1}$$

False positive (FP) rate – processes the amount of authentic cases which are erroneously identified as phishing assaults in connection to all current real occurrences as exhibited in Eq. 2.

$$\mathrm{FP} = \frac{N_{\mathrm{L->P}}}{N_{\mathrm{L->L}} + N_{\mathrm{L->P}}} \tag{2}$$

True negative (TN) rate – processes the amount of accurately identified real cases in connection to all current real occasions as exhibited in Eq. 3.

$$\mathrm{TN} = \frac{N_{\mathrm{L->L}}}{N_{\mathrm{L->L}} + N_{\mathrm{L->P}}} \tag{3}$$

False negative (FN) rate – processes the amount of phishing assaults which are mistakenly identified real in connection to all the current phishing assaults as exhibited in Eq. 4.

$$\mathrm{FN} = \frac{N_{\mathrm{P->L}}}{N_{\mathrm{P->P}} + N_{\mathrm{P->L}}} \tag{4}$$

Precision (P) – processes the amount of accurately identified phishing assaults in connection to all the occasions that are distinguished as phishing as exhibited in Eq. 5.

$$P = \frac{N_{\mathrm{P->P}}}{N_{\mathrm{L->P}} + N_{\mathrm{P->P}}} \tag{5}$$

Recall (R) – same as TP. Equation 6 reveals the details.

$$R = \mathrm{TP} \tag{6}$$

f_1 score – is symphonious mean among P and R as showed in Eq. 7.

$$f_1 = \frac{2PR}{P+R}. \tag{7}$$

Accuracy (ACC) – processes the general amount of accurately identified phishing and authentic examples in connection to all occasions. See Eq. 8 for subtleties

$$\text{ACC} = \frac{N_{\text{L->L}} + N_{\text{P->P}}}{N_{\text{L->L}} + N_{\text{L->P}} + N_{\text{P->L}} + N_{\text{P->P}}} \quad (8)$$

Weighted error (W_{ERR}) – measures by and large weighted degree of erroneously identified phishing and authentic cases in connection to all cases.

Discovery of Phishing Bouts: The Humanoid Factor

While phishing assaults endeavor to exploit the unpracticed clients, a conspicuous arrangement is teaching the clients, which would thus diminish their helplessness to falling casualties of phishing assaults. Various client-preparing tactics have been projected all through the previous years.

A. *Phishing Sufferers*

Julie S. Downs et al. (2007) studied 232 PC clients to think about the various principles that can foresee the helplessness of the client to tumble unfortunate casualties for phishing messages. The review was framed in a pretend where every client relied upon to investigate messages just as responding to various inquiries. The result of the investigation was that the individuals who had a decent learning on the "phishing" definition were remarkably less inclined to succumb to phishing messages, while information about different territories, for example, threats, spyware, and infections, had not helped in decreasing defenselessness to phishing messages.

Another examination that confirms the investigation in (Downs et al. 2007) was done by Huajun Huang et al. (2009), who presumed that the essential motives that make innovation clients tumble as unfortunate casualties for phishing assaults seem to be:

- Clients overlook detached admonitions (e.g., toolbar markers).
- Countless can't separate among phishing and genuine destinations, regardless of whether they are informed that their capacity is being tried.

A statistic examination made by Steve Sheng et al. (2010) demonstrates various roundabout attributes that connect among unfortunate casualties and their defenselessness to phishing assaults. As indicated by their examination, sexual orientation and age unequivocally connect with phishing vulnerability. They presume that:

Females will in general snap on email interfaces more regularly than guys.
Individuals somewhere in the range of 18–25 years of age were significantly more likable to succumb to phishing assaults than other age gatherings.

B. *User-Phishing Communication Model*

Xun Don et al. (2008) portrayed the first visual client phishing communication demonstration. The prototype portrays client cooperation from the basic leadership perspective, which begins the minute a client visualizes phishing substance and closures when most of the client activities are finished. The objective is helping the way toward relieving phishing assaults by seeing client's collaboration on phishing content.

The decision-making process involves the below inputs:

- Exterior information: knowledge through the user interface (UI) or master guidance.
- Information and context: the client's present comprehension of the world that works after some time.
- Expectation: clients have desires dependent on their comprehension and their activity results.

Phishers can just alter the choice procedure of clients through giving outside data through the UI. The UI gives two informational indexes:

- Metadata: URL in web browser address bars, or it may be an email address.
- Content data: email or website data.

Phishing assaults prosper only when a phishing assault persuades the client's real meta-information and content information. Meta-information is utilized by the clients to decide upon the authenticity of the emails. To trap the clients, additionally, phishers parody meta-information of the clients. The answer for meta-information honesty issue isn't through client instruction or mindfulness as it is very difficult for clients to approve whether the source IP address is real in the event that the space name was satirize. Clients ought not be relied upon to approve the meta-information as it is somewhat a framework structure or usage issue. Then again, through social designing, phishers make persuading and authentic-looking substance information. A typical answer for this is client mindfulness.

C. *Administration Policies*

Associations ought to take severe strategies against the appropriation of confidential information over email, SMS, or VoIP, in addition to client's awareness about the strategies. Clients that are influenced mindful of this to have a higher possibility of distinguishing irregularities inside a phishing message if the substance information, for instance, requested confidential data. Specialist organizations ought to likewise entirely uphold their strategies against unlawful utilization of their administrations. Many facilitating suppliers bring administrations down in situations where they are manhandled. As inquired about by T. Moore and R. Clayton in (2007), administration overthrow is progressively regular in manner in which security issues are dealt with.

D. *Inactive and Lively Warnings*

UIs can demonstrate safety admonitions dependent on activated activities, for example, seeing phishing site pages, as usually sent by lot of Internet browsers. Displaying the security breaches to the clients can be done in two ways:

- Inactive/passive warnings – the notice does not hinder the content area and empowers the client to see both the content.
- Lively/active warnings – the notice obstructs the content information, which restricts the client from reviewing the content information.

E. *Educational Media*

E-administrations, for example, e-banking, frequently impart occasional instructive messages to caution their customers of potential phishing dangers. The messages are regularly conveyed by means of SMSs and messages. As indicated by an examination led by Kumaraguru et al. (2007), occasional safety sees are insufficient, and in spite of the fact that they may increase the information of end clients, the intermittent notification neglects to alter their conduct. On the other hand, Kumaraguru et al. (2007) suggests and assesses the structure of an option occasional technique to send instructive notification that is installed into the day-by-day action of the end clients. The investigation demonstrates that inserted preparing frameworks are more powerful than occasional security notes.

The proposed framework works as follows:

- The email executive readies various phony phishing messages.
- The phishing messages are conveyed to the people in question. No alerts appeared at this stage.
- When the injured individual connects with a phishing electronic message, for example, by tapping on a phishing join, the client is then demonstrated a safety caution showing the dangers of phishing assaults.

Phishing Recognition by Blacklists

Blacklists are much of the time refreshed arrangements of recently recognized phishing URLs or watchwords. Whitelists, then again, are inverse and could be utilized to decrease FP rates. Blacklists don't give security against party time phishing assaults as a site should be recently recognized first so as to be boycotted. Be that as it may, blacklists by and large have lesser FP rates than heuristics (Sheng et al. 2009). As concentrated in (Sheng et al. 2009), blacklists are observed to be incapable against party time phishing assaults and had the capacity to identify just 20% of them. The examination (Sheng et al. 2009) likewise demonstrates that 48–87% of URL phishing were blacklisted in 12 hours. This postponement is a great issue as 64% of phishing efforts stopped inside the first 2 hours.

A. *Browse Safe API by Google*

Browse safe API by Google empowers customer requests to approve a provided URL appear in blacklists, and the lists are always refreshed by Google (2011). Though this is a trial, this method is utilized by Chrome and Firefox. The present execution of convention is given by Google and just comprises two blacklists named goog-phishshavar and goog-malware-shavar, for phishing and malware separately. The customer application downloads records from suppliers and keeps up its cleanness by steady updates that comprise added substance and subtractive lumps. The convention pursues a draw demonstration that necessitates the customer to interface with the server at certain number time interims estimated in minutes (tm).

B. *Blacklist Using DNS*

DNS Blacklist (DNSBL) suppliers utilize the standard DNS convention. Because of its utilization of the DNS specification, any standard DNS server could go used as DNSBL. Nonetheless, since the quantity of recorded sections is huge, a server that isn't streamlined for dealing with a lot of DNS or text resource records (RRs) confront execution and asset strains. Rbldnsd (Anti-Phishing Working Group 2010) is a quick DNS server, structured precisely to deal with huge RR appropriate for DNSBL obligations.

At the point when a message transfer agent (MTA) sets up an inbound SMTP association, it confirms whether interfacing source is recorded in phishing blacklists (Anti-Phishing Working Group 2011a), by transferring a DNS ARR inquiry to a DNSBL server on the associating IP address. Contingent upon the DNSBL server, the blacklisting cause would be questioned by means of DNS TXT RR inquiry. Notwithstanding, if no DNS discovers a RR, MTA considers that source IP address is not blacklisted.

C. *Phish.net: Prognostic Blacklisting*

Phish.net (Prakash et al. 2010) discourses the careful competition constraint found in boycotts. On account of this, Phish.net forms boycotted URLs (guardians) from which different varieties of similar URL (youngsters) are created through five diverse varieties of URL heuristics, and those are recorded beneath:

Supplant top-level domains: Individual URL will split into 3210 varieties with an alternate top-level domains.

Directory arrangement similarity: If various phishing URLs have comparable catalog arrangements with tiny varieties, different kid's URLs are made to collect contrasts over all of the assault URLs that include comparative registry structures.

IP address likeness: URLs that include comparative index arrangement yet extraordinary area names are considered as a match on the off chance that points to a similar IP address.

Query string replacement: Alike to "directory arrangement similarity"; then again, actually, it splits numerous varieties of URL with various question strings.

Brand term likeness: Numerous kid's URLs with same URL, however, with various brand terms. For instance, http://www.ayc.com/ on the web/paypal.htmls would likewise fork http://www.ayc.com/ on the web/ebay.htmls as a kid (as they are regularly assaulted).

A significant number of the forked kid's URLs might not exist or might be guiltless unimportant pages. In order to expel/filter nonexistent kid's URLs, accompanying checks are performed:

DNS enquiry: Existence of the domain name.
TCP join: Was the HTTP server utilized by the resolved name?
HTTP header answer: Page existence report (HTTP 404 Not Found Error).

Content resemblance: Is the content alike to parent phishing bout? On the off chance that the page is unique, it may be an honest page. An outside tool (Anti-Phishing Working Group 2011b) was utilized to gauge the substance comparability.

D. *Robotized Discrete Whitelist*

Robotized Discrete Whitelist (RIWL) (Cao et al. 2008) keeps up a whitelist of highlights depicting trustworthy Login UIs (user interfaces) where the client presented his/her qualifications. Each Login UI will provide a notice aside from whenever trusted. When a Login UI is believable, the highlights of it will be put away in a whitelist locally.

There are two principal modules of RIWL:

Whitelist – a rundown of trustworthy Login UIs. Its goal is smothering admonitions with respect to Login UIs that are trustworthy. So as to believe a suspected Login UI, the Login UIs are spoken to as an element vector and afterward looked at the whitelist against those element vectors. On the off chance that the Login UI highlights a suspicious page that doesn't coordinate with the majority of the highlights in any whitelisted Login UI, the presumed page is thought as not trustworthy, and admonitions are exhibited to the end client.

Computerized whitelist maintainer – the classifier determines whether a suspicious Login UI is to be introduced in the whitelist. The number of times the end client signs in using the Login UI is recorded as the value of whitelist maintainer. If the value is more, then the Login UI is accepted as trustworthy. Thus, the computerized whitelist maintainer checks for effectiveness of the Login UIs. After that, the classifier continuously observes the number of fruitful logins per suspicious Login UI, and if the effective login attempts surpass a prescribed range, it is whitelisted. The contribution of the classifier is that it portrays the substance of a suspicious Login UI (not in the whitelist) and yields a result of whether to confide in the Login UI.

Phishing Discovery: Heuristic Approach

Programming can be introduced on the client/server to review payloads of different conventions through diverse calculations. Conventions could be HTTP, SMTP, or any self-assertive convention. Calculations could be any system to distinguish or forestall phishing assaults. Phishing heuristics exist in phishing assaults in all actuality; anyway, the attributes are not ensured to dependably occur in such assaults. Various heuristic tests are identified, and it may be conceivable to distinguish the time of phishing assaults (e.g., assaults that are not gotten beforehand), which is a leverage against boycotts (since boycotts need definite matches, accurate assaults should be watched first so as to boycott them). Nonetheless, such summed up heuristics additionally risk misclassifying real substance (e.g., real messages or sites).

Presently, significant Internet surfers and email customers are worked with phishing security instruments, for example, heuristic tests that go for recognizing phishing assaults. The customers incorporate Mozilla Firefox, Google Chrome, Internet Explorer, UC Browser, and MS Outlook. Likewise, phishing location heuristics are incorporated into antiviruses, like ClamAV10.

A. *SpoofGuard*

SpoofGuard (Chou et al. 2004), an Internet browser module created at Stanford University, distinguishes HTTP-based phishing endeavors as an Internet browsing toolbar, by determining certain abnormalities in the HTML content with respect to a defined edge esteem.

B. *Cooperative Intrusion Discovery*

Numerous phishing identification and counteractive action components depend on identifying the source IP of the aggressor. Quick flux (Holz et al. 2008), then again, empowers assailants to much of the time alter their IPs. Lot of contaminated hosts are available that leads the clients to phishing sites. Different intermediaries during browsing hand off the traffic to a primary phishing site to get the substance from (otherwise called mother ship). A proposed answer for this is using cooperative intrusion discovery system (CIDS) to trade phishing interrelated information among various intrusion discovery systems (IDSs). The circulated framework ought to be introduced internationally. Neighbor CIDS scan its local DNS reserve to identify the DNS centers with a large number of DNS ARR grouped with low TTL values. This information is spread worldwide for CIDS. Every beneficiary CIDS in turn scan the exposed IP addresses (which contaminates clients much of the time). Thus, by continuous monitor of incoming and outgoing associations of suspicious IPs, the source phishing site is determined.

C. *PhishGuard*

The work in (Likarish et al. 2008) puts together its assurance against phishing with respect to the possibility that phishing sites don't frequently check client qualifications, but save them for future use. The creators in (Likarish et al. 2008) recognize that, later on, phishing locales would be increasingly refined and queued (or passage) their yield from authentic destinations, going about as a man in the centre assault, which would thus result in legitimate achievement or disappointment login alerts (as correspondence is basically channeled forward and backward). Be that as it may, the paper expresses that such use isn't normal yet and basically concentrates its location on non-channeled phishing endeavors. PhishGuard's execution in (Likarish et al. 2008) is a proof of idea that just identifies phishing assaults dependent on testing HTTP digest verifications. In any case, coordinating other verification systems, for example, through HTML structure accommodation, is conceivable. PhishGuard pursues the accompanying strides to test a suspected page.

Phishing Detection by Visual Similarity

This area traces various proposed arrangements that endeavor to identify phishing assaults dependent on their visual appearance, rather than dissecting the fundamental source code organize level data.

A. *Discriminative Feature Classification*

Dissimilar to other enemy of phishing instruments, the projected arrangement in (Chen et al. 2009) reveals phishing identification dependent on the content demonstration, rather than content code. At the end of the day, this phishing location component is skeptic to the hidden code or innovation that reduces the last visual yield to client's eyes.

B. *Visual Similarity-Based Discovery Without Target Site Information*

The objective of the proposed method in (Hara et al. 2009) is identifying phishing locales dependent on visual closeness without using whitelisting photos of every single authentic site and depends on the way that most phishing sites go for being outwardly like their objective sites (e.g., PayPal phishing sites mean to look outwardly like the genuine PayPal site so as to augment their odds of influencing more unfortunate casualties).

Phishing Detection by Data Mining

Like heuristic tests, machine learning (ML)-based strategies can moderate party time phishing assaults, which makes them worthwhile when contrasted and boycotted. Strikingly, ML procedures are likewise fit for developing individual classification models by examining substantial arrangements of information. This hoists the necessity of physically making heuristic tests as machine learning calculations by which individual models can be found. At the end of the day, ML strategies have the accompanying focal points over heuristic tests:

- Regardless of the confused idea of antagonistic assaults, it is conceivable to develop powerful classification models when expansive informational collection tests are accessible, without the necessity of physically dissecting information to find complex connections.
- As phishing efforts develop, machine learning classifiers can naturally advance by means of fortification learning. On the other hand, it is likewise conceivable to intermittently build more up-to-date classification prototypes by basically reeducating the student with refreshed example of informational indexes.

ML-based enemy of phishing classifiers in the writing, for example, those exhibited in (Whittaker et al. 2010; Bergholz et al. 2010), demonstrates that it is conceivable to accomplish under 1.5% false positive and over 99.0% true positive rates. As indicated by the overviewed writing, machine learning-based classifiers are the main classifiers that accomplished such high classification precision in identifying the phishing assaults.

3 Proposed Approach

The deliberate structure of the model proposed as outlined in Fig. 2 contains the arrangement of authentic URLs and phishing URLs. Pre-handling and standardization forms are completed in the element extraction stage. Determination of high-

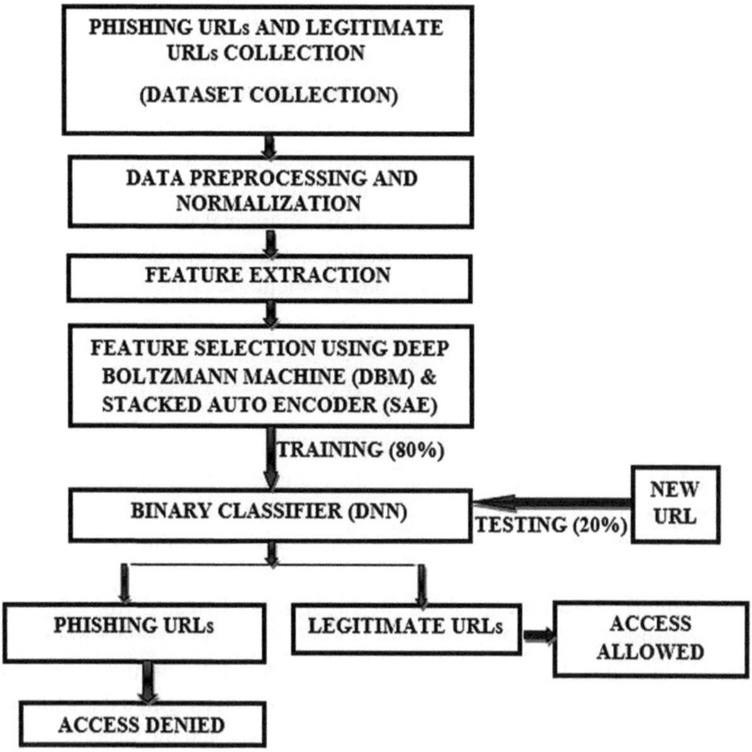

Fig. 2 Deliberate structure of the proposed model

lights is performed utilizing deep Boltzmann machine (DBM) and stacked auto-encoder (SAE) in highlight choice stage. Double grouping of authentic and phishing URLs is completed utilizing DNN.

3.1 *URL Collection*

We gathered URLs of amiable sites from The Web Information Company (Brewster Kahle et al. 1996), DMOZ Open Directory Project (Rich Skrenta et al. 1998), and individual Internet browser history. The phishing URLs were gathered from PhishTank (David Ulevitch 2006). The informational index comprises of 17,000 phishing URLs and 20,000 legitimate URLs. We got PageRank (Ian Rogers 2002) of 240 legitimate sites and 240 phishing sites by checking PageRank independently at PR Checker (Roger). We gathered WHOIS (WHOIS look up) data of 240 considerate sites and 240 phishing sites.

3.2 Preprocessing and Feature Extraction

In this stage, the unstructured data about the URL (e.g., literary portrayal) is fittingly arranged and changed over to a numerical vector with the goal that it very well may be nourished into pre-preparing calculation. Pre-preparing has been completed to evacuate the loud and missing qualities from the numerical vectors. Plus, Z-score standardization is utilized on the numerical vector to standardize the qualities in the range between 0 and 1 (Selvaganapathy et al. 2018). The accompanying condition is utilized for standardization practice:

$$z = \frac{x - \mu}{\sigma} \tag{9}$$

where x is the numeric value for each sample of the feature, μ is the mean of the feature, and σ is the standard deviation of the feature. The absolute value of z represents the distance between the numeric value and the mean of the feature in units of the standard deviation. z is undesirable when the numeric value is under the mean and optimistic when beyond.

These numerical vectors speak to the highlights extricated from the gathered URLs. The gathered highlights are arranged into two kinds: lexical highlights and host-based highlights. In general, lexical sorts speak to 62% of highlights and host-based highlights speak to 38%. Phishing URLs can be investigated dependent on the lexical highlights and host-based highlights of the URL, and the structures appeared in Fig. 3. Thus, 30 highlights are removed from the gathered URLs which are recorded in the Table 2.

3.3 Pre-training of the Features and Feature Selection

3.3.1 Deep Boltzmann Machine-Based Feature Selection

A deep Boltzmann machine is a system of symmetrically coupled stochastic binary units (Salakhutdinov and Larochelle 2010). It contains an arrangement of visible units $v \in \{0, 1\}^D$ and a grouping of layers of hidden units $h^1 \in \{0, 1\}$ F^1, $h^2 \in \{0, 1\}$ F^2, $h^L \in \{0, 1\}$ F^L. There are associations just between hidden units in contiguous layers and also between the visible units and the hidden units in the primary hidden layer. The energy of the state {v, h} is characterized as:

$$E(\mathrm{v,h};\theta) = -\mathrm{v}^{\mathrm{T}}\mathrm{W}^1\mathrm{h}^1 - \mathrm{h}^{1\mathrm{T}}\mathrm{W}^2\mathrm{h}^2 - \mathrm{h}^{2\mathrm{T}}\mathrm{W}^3\mathrm{h}^3, \tag{10}$$

where $h = \{h^1, h^2, h^3\}$ are the set of hidden units and $\theta = \{W^1, W^2, W^3\}$ are the model parameters, representing visible-to-hidden and hidden-to-hidden symmetric interaction terms (Salakhutdinov and Larochelle 2010). The likelihood that the model assigns to a visible vector v is:

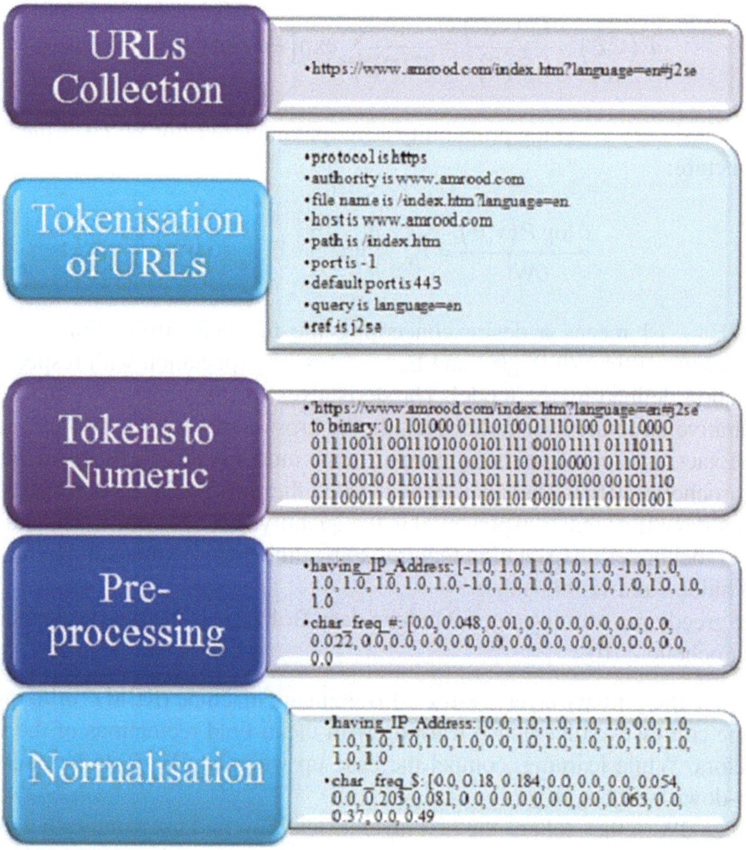

Fig. 3 URL tokenization and feature extraction

Table 2 List of extracted features from phishing URLs

Phishing URL features (30 features)	having_IP_Address, URL_Length Shortining_Service, having_At_Symbol double_slash_redirecting, Prefix_Suffix having_Sub_Domain, SSLfinal_State Domain_registeration_length, Favicon Port, HTTPS_token, Request_URL, URL_of_Anchor, Links_in_tags, SFH Submitting_to_email, Abnormal_URL Redirect, on_mouseover, RightClick popUpWidnow, Iframe, age_of_domain DNSRecord, web_traffic, Page_Rank Google_Index, Links_pointing_to_page, Statistical_report

$$P(\mathrm{v};\theta)=\frac{P^{*}(\mathrm{v};\theta)}{\mathcal{Z}(\theta)}=\frac{1}{\mathcal{Z}(\theta)}\sum_{\mathrm{h}}\exp\left(-E\left(\mathrm{v},\mathrm{h}^{1},\mathrm{h}^{2},\mathrm{h}^{3};\theta\right)\right). \tag{11}$$

The subsidiary of the log probability as for parameter vector W^1 takes the following structure:

$$\frac{\partial \log P(\mathrm{v};\theta)}{\partial \mathrm{W}^{1}}=\mathrm{E}_{P_{data}}\left[\mathrm{vh}^{1^{T}}\right]-\mathrm{E}_{P_{model}}\left[\mathrm{vh}^{1^{T}}\right], \tag{12}$$

where E_{pdata} [·] means a desire concerning the finished information circulation $P_{data}(h, v; \theta) = P(h|v; \theta)\ P_{data}(v)$ and E_{pmodel} [·] is an expectation with respect to the distribution defined by the model. The derivatives for parameters W^2 and W^3 take comparative structures yet rather include the cross items h^1h^{2T} and h^2h^{3T}, respectively. Exact maximum likelihood learning in this model is intractable (Salakhutdinov and Larochelle 2010). The right estimation of the information subordinate desire requires some time that is exponential in the quantity of hidden units, while the right count of the model's expectation requires some time that is exponential in the quantity of hidden and visible units.

The greedy pre-training algorithm for a deep Boltzmann machine (Salakhutdinov and Larochelle 2010):

1. Prepare the primary layer "restricted Boltzmann machine (RBM)" utilizing one-stage contrastive divergence learning with mean-field recreations of the visible vectors. While learning, compel the base up weights, $2W^{(1)}$, to be double the top-down weights, $W^{(1)}$.
2. Freeze $2W^{(1)}$ that defines the first layer of features, and use samples $h^{(1)}$ from P ($h^{(1)}$ |v; $2W^{(1)}$ as the information for preparing the second RBM. This is an appropriate RBM with weights $2W^{(2)}$ that are of a similar size in both directions. It is also trained using one-step contrastive divergence learning with mean-field reconstructions of its visible vectors.
3. Freeze $2W^{(2)}$ that defines the second layer of features, and use the samples $h^{(2)}$ from P ($h^{(2)}$ |v; $2W^{(1)}$, $2W^{(2)}$) as the data for training the next RBM in the same way as the previous one.
4. Continue iteratively up to layer $L - 1$.
5. Train the top-level RBM using one-step contrastive divergence learning with mean-field reconstructions of its visible vectors. During the learning, compel the base up weights, $W^{(L)}$, to be half the top-down weights, $2W^{(L)}$.
6. Utilize the weights $\{W^{(1)}, W^{(2)}, W^{(3)}, \ldots, W^{(L)}\}$ to create a deep Boltzmann machine.

This phase comprises pre-training a pile of restricted Boltzmann machines (RBMs) with the unsupervised, greedy contrastive divergence (CD) algorithm by randomly initializing the feature values to the nodes of input layer.

The pre-prepared component estimations of RBM are used as the report information for preparing the accompanying RBM in the stack, with the result that the nonlinear, high-dimensional unlabeled information is tested as best, low-dimensional portrayal of the highlights. Totally, after the pre-preparing, a DBM is produced by unrolling the RBMs as appeared in the calculation under area III, and a short time later, the DBM is calibrated using backspread of mistake subordinates.

3.3.2 Deep Stacked Auto-Encoder-Based Feature Selection

A basic auto-encoder (AE) is a deep learning architecture model in which an original URL feature at the input is reconstructed at the output going through an intermediate layer with reduced number of hidden nodes (Yoshua and Pascal 2007). The AE model tries to learn deep and abstract features in those reduced hidden nodes, so a reconstruction is feasible from them.

The AE training consists of reproducing input URL features at the output of the model, so internal units are able to provide the original information. Ensuring a proper reconstruction, the values in the layer can then be employed as new reduced features representing the original URL features u. The AE satisfies:

$$\begin{aligned} n &= f\left(w_n \mathbf{u} + b_n\right) \\ m &= f\left(w_m \mathbf{y} + b_m\right) \end{aligned} \tag{13}$$

where the interval variable is obtained from u by the weights w_n and common bias b_n and the reconstructed signal supposed to match u is obtained directly from the layer output y by w_m and b_m; f represents the activation function, which introduces the nonlinearity in the network. To train the AE and determine the optimized parameters, the error between u and m needs to be minimized, i.e.,

$$\underset{w_n w_m b_n b_m}{\arg\min}\left[\text{error}\left(\mathbf{u},\mathbf{m}\right)\right] \tag{14}$$

SAEs can be defined by expanding this concept and simply introducing several layers between the input and the output. Therefore, final features are obtained through progressive abstraction levels.

The preparation procedure in SAEs comprises of an iterative refresh of the various interior coefficients w and b, a refresh by which the mistake between the information pixel and the reproduced one at the yield of the system is logically decreased until the point when it is underneath some esteem or edge. A viable preparing converts into a decreased blunder as communicated in condition, which guarantees suitable inward highlights. In the proposed framework, a profound neural system for pre-preparing process contains an info layer, six shrouded layers, and a yield layer. The quantities of information highlights are 30. The shrouded layers contain five neurons each. The aftereffects of the centre layers have been fell to frame a SAE conspire for highlight decrease same. Diverse arrangement of highlights is removed by differing the quantity of concealed layers utilized. The location exactness of the distinctive highlights extricated is examined in Sect. 4. The element vector got at the yield layer has been diminished to a component of nine highlights. The separated highlights are approved dependent on the precision in identification of URLs.

3.4 Detection and Classification

Prior to recognizing and grouping the phishing and benevolent URLs utilizing paired classifier DNN, the preparation test (80%) and testing test (20%) are separated from the pre-prepared component esteems. The preparation of DNN classifier display is completed utilizing the separated preparing dataset for the location of the phishing and real URLs. At that point, the prepared DNN twofold grouping calculation is utilized to order testing URL tests.

4 Experimental Results

4.1 Data Sets

4.1.1 Conventional Dataset

Alexa dataset (The Web Information Company): Alexa dataset is utilized as a benchmark dataset of amiable and ordinary sites. Alexa is a business organization that plays out the errand of web traffic information investigation over the web. It acquires clients' perusing designs from different sources and basically investigates them for web traffic announcing and positioning of URLs on the Internet. The rankings given by Alexa are utilized by specialists to collect a lot of exceptionally positioned sites as a typical dataset for testing and recognizing TNR. Alexa gives the dataset of typical sites as a crude content record in which each line makes reference to the position of a site and its area name in climbing request.

4.1.2 Phishing Dataset

PhishTank dataset (PhishTank): PhishTank dataset is utilized as a benchmark dataset of phishing sites. PhishTank is a community-based framework that confirms phishing sites. An assortment of clients and outsiders submit suspected phishing destinations that are in the long run casted a ballot on by a lot of clients for their accessibility as a substantial phish. PhishTank therefore gives a real-time dataset of phishing sites. Phishing sites given by PhishTank are utilized by specialists to make a dataset of phishing sites for testing and distinguishing the TPR. PhishTank dataset is accessible in CSV record design, and each line of the document contains subtleties of an interesting phish announced over PhishTank. The subtleties incorporate phish ID, phish URL, phish detail URL, accommodation time, checked status, confirmation time, online status, and target URL.

4.2 *Effect of Dynamic Feature Extraction and Pre-training*

As of late, deep learning has been demonstrating promising outcomes in different artificial intelligence applications like picture acknowledgment, normal dialect handling, dialect displaying, and neural machine interpretation (Sewak et al. 2018). In this way, in this paper, we thought about the adequacy of a deep learning architectures called deep Boltzmann machine, stacked auto-encoder, and deep neural network (DNN) with the established machine learning calculations for the phishing URL grouping. We utilized distinctive profound learning-based element choice/extraction methods and accomplished a greatly improved precision (~94%) utilizing deep neural network with a shockingly better FPR (~5%). Our investigation demonstrates that deep learning-based designs, for example, auto-encoders for highlight extraction joining with deep neural networks for the arrangement, give measurably better exactness when contrasted with condition of workmanship machine learning method mixes. The list of capabilities will powerfully change dependent on the example input URL. As the list of capabilities changes powerfully, it is unimaginable to expect to adhere to an explicit arrangement of chose highlights for recognizing all the assault types. The arrangement of highlights that results after pre-preparing includes roughly nine highlights.

4.3 *Experimental Results on Feature Selection*

The parameter that is utilized for the approval of the highlights chosen is the rate of misclassification mistake, and a while later, the DBM and SAE are adjusted by using backspread of blunder subordinates. The misclassification blunder was determined utilizing the equation:

$$\text{Misclassification error} = \left(\frac{\text{Number of samples incorrectly classified}}{\text{Total number of samples}} \right) * 100 \qquad (15)$$

Figure 4 delineates distinctive arrangement of highlights that were extricated by changing the quantity of shrouded layers in both deep Boltzmann machine and auto-encoder. At that point, these capabilities were given to the location stage, and the precision was determined. The list of capabilities with the better precision was viewed as the most important list of capabilities. Consequently, the auto-encoder gives the most applicable highlights.

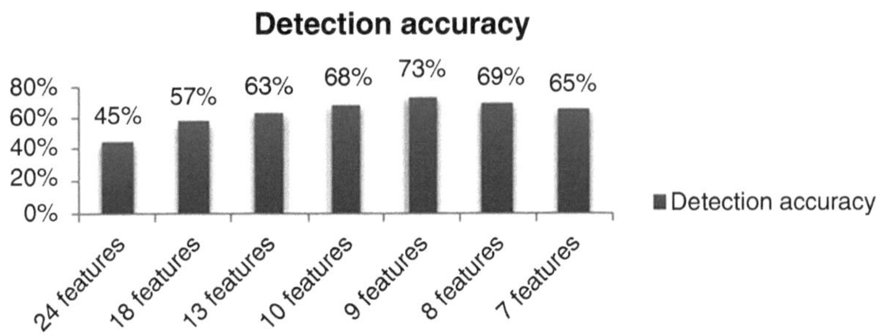

Fig. 4 Detection accuracy for different sets of reduced features

The performance of DBM and SAE as a feature reduction technique is compared with other feature reduction techniques such as principal component analysis (PCA), attribute selection, gain ratio, and Chi-square which is shown in Table 3, and it is observed in Fig. 5 that SAE provides better results (Selvaganapathy et al. 2018).

4.4 Evaluation Measures

Precision is the rate of right expectations that the model will accomplish when contrasted with the real groupings in the dataset. Then again, precision and review are two assessment systems, which determine depending on disarray lattice as appeared in Table 4 and figured by Eqs. 16, 17 and 18:

$$\text{Precision} = \text{TP} / (\text{TP} + \text{FP}) \tag{16}$$

$$\text{Recall} = \text{TP} / (\text{TP} + \text{FN}) \tag{17}$$

$$\text{Accuracy} = (\text{TP} + \text{TN}) / (\text{TP} + \text{FP} + \text{TN} + \text{FN}) \tag{18}$$

where:

True positive (TP): the number of correct detected phishing URLs
True negative (TN): the number of legitimate emails that was detected as legitimate URLs
False positive (FP): the number of legitimate emails that was detected as phishing URLs
False negative (FN): the number of phishing emails that was detected as legitimate URLs

Table 3 Accuracy comparison of DBM and SAE with other feature reduction techniques

Feature reduction methods	Accuracy
PCA	68%
Attribute selection	63%
Gain ratio	65%
Chi-square	66%
DBM	70%
SAE	73%

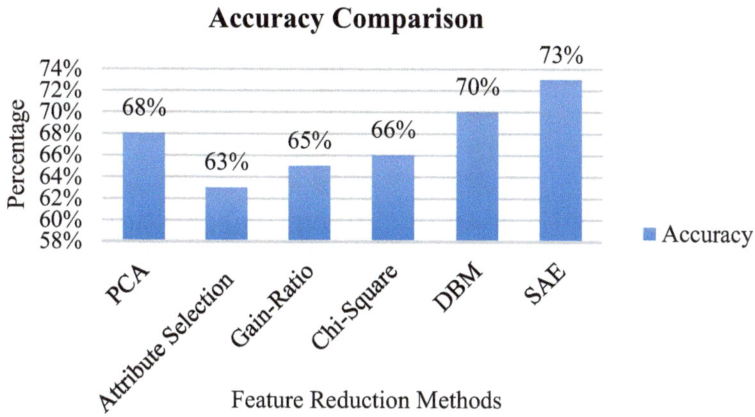

Fig. 5 Performance comparison of feature reduction methods

Table 4 Confusion matrix

	Classified phishing URL	Classified legitimate URL
Actual phishing URL	TP	FN
Actual legitimate URL	FP	TN

4.5 Comparison of Deep Learning Results with Machine Learning Techniques

The analysis is performed in the PC with Intel(R) Core(TM) i5 1.80 GHz processor of 4 GB RAM memory, and the program is coded in RStudio programming (Selvaganapathy et al. 2018). Table 5 demonstrates the execution aftereffects of proposed technique, in which DNN calculation is contrasted and machine learning calculations utilized in the current framework (James et al. 2013). By utilizing lexical and content-based highlights, we had the capacity to accomplish a higher detection accuracy/success rate of 94.73% than support vector machine (SVM), Naïve Bayes (NB), Regression Tree, and K-nearest neighbor (KNN).

Table 5 Performance comparison of DNN with other machine learning algorithms

Methods	Accuracy	TP rate	FP rate
DNN	94.73	97.62	5.27
SVM	87.65	90.03	12.35
Naive Bayes	74.20	95.47	25.80
Regression Tree	91.08	95.47	8.92
KNN	79.55	80.14	20.45

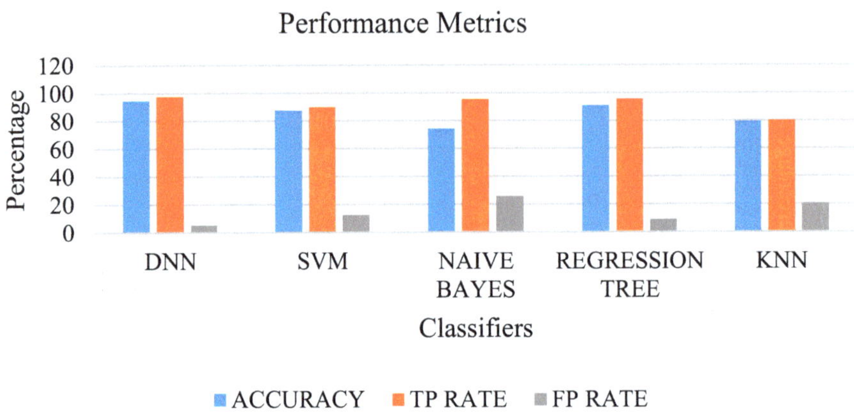

Fig. 6 Detection accuracy of various models

4.5.1 R and RStudio (Statistical & Qualitative Data Analysis Software 2019)

R is a free, open-source programming program for measurable examination, in view of the S dialect. RStudio is a free, open-source IDE (incorporated improvement condition) for R. (You should introduce R before you can introduce RStudio.) Its interface is composed so the client can unmistakably see charts, information tables, R code, and yield all in the meantime. It additionally offers an Import Wizard-like component that enables clients to import CSV, Excel, SAS (∗.sas7bdat), SPSS (∗.sav), and Stata (∗.dta) documents into R without composing the code to do as such.

Figure 6 demonstrates a correlation of discovery exactness parameters of DNN, SVM, Naïve Bayes, Regression Tree, and KNN classifiers (James et al. 2013). Additionally, Fig. 6 demonstrates a correlation of TP rate and FP rate of DNN, SVM, Naïve Bayes, Regression Tree, and KNN classifiers separately. Investigating the information demonstrates that DNN paired classifier performs well with higher true positive (TP) rate of 97.62% and lower false positive (FP) rate of 5.27%.

5 Conclusion

The anticipated framework demonstrates that the deep learning techniques DBM and SAE can be proficiently and adequately utilized in the phishing URL recognition and categorization than other AI calculations. SAE with three concealed layers utilized in the anticipated framework indicates improved great dimensional depiction of reduced feature set than DBM. The DNN parallel classification technique has decreased the bogus positive rates than other AI calculations in distinguishing the phishing URLs. The trouble in the anticipated strategy is preparing the whole DBM model which is tedious with directionless associations between the layers. In upcoming research, the previously mentioned deficiency of the DBM model can be beaten by utilizing deep belief network (DBN) which utilizes coordinated associations of the hubs between the layers.

References

Anti-Phishing Working Group (APWG). (2010). *Phishing activity trends report second half 2010*, http://apwg.org/reports/apwg report h2 2010.pdf. Accessed Dec 2011.

Anti-Phishing Working Group (APWG). (2011a). *Phishing activity trends report first half 2011*, http://apwg.org/reports/apwg trends reporth12011.pdf. Accessed Dec 2011.

Anti-Phishing Working Group (APWG). (2011b). *Phishing activity trends report second half 2011*, http://apwg.org/reports/apwg trends reporth22011.pdf. Accessed July 2012.

Bergholz, A., De Beer, J., Glahn, S., Moens, M.-F., Paaß, G., & Strobel, S. (2010). New filtering approaches for phishing email. *Journal of Computer Security, 18*, 7–35.

Brewster, K., & Bruce, G. (1996). The Web Information Company., www.alexa.com

Cao, Y., Han, W., & Le, Y. (2008). Anti-phishing based on automated individual white-list. In *DIM '08: Proceedings of the 4th ACM workshop on digital identity management* (pp. 51–60). New York: ACM.

Chen, K.-T., Chen, J.-Y., Huang, C.-R., & Chen, C.-S. (2009). *Fighting phishing with discriminative keypoint features. Internet Computing, IEEE, 13*(3), 56–63.

Chou, N., Ledesma, R., Teraguchi, Y., & Mitchell, J. C. (2004). Client-side defense against web-based identity theft. In *NDSS*. The Internet Society.

David Ulevitch. (2006). PhishTank. http://www.phishtank.com

Dong, X., Clark, J., & Jacob, J. (2008). *Modelling user-phishing interaction in Human System Interactions*, Conference on, 2008, May, pp. 627–632.

Downs, J. S., Holbrook, M., & Cranor, L. F. (2007). Behavioral response to phishing risk. In *Proceedings of the anti-phishing working groups 2nd annual eCrime researchers summit, ser. eCrime '07* (pp. 37–44). New York: ACM.

Google. Google safe browsing API, http://code.google.com/apis/safebrowsing/. Accessed Oct 2011.

Hara, M., Yamada, A., & Miyake, Y. (2009). Visual similarity-based phishing detection without victim site information. In *IEEE symposium on computational intelligence in cyber security, 2009. CICS '09* (pp. 30–36).

Holz, T., Gorecki, C., Rieck, K., & Freiling, F. C. (2008). Measuring and detecting fast-flux service networks. In *Proceedings of the network and distributed system security symposium (NDSS)*.

Huang, H., Tan, J., & Liu, L. (2009). Countermeasure techniques for deceptive phishing attack. In *International conference on new trends in information and service science, 2009. NISS '09* (pp. 636–641).

Ian Rogers. (2002). Google Page Rank – Whitepaper, http://www.srigroane.net/google-page-rank/, http://www.prchecker.info/check_page_rank.php

James, J., Sandhya, L., & Thomas, C. (2013). *Detection of phishing URLs using machine learning techniques*. IEEE international conference on control communication and computing (ICCC).

Knickerbocker, P., Yu, D., & Li, J. (2009). Humboldt: A distributed phishing disruption system. In *eCrime researchers summit* (pp. 1–12).

Krebs, B. (2011). *HBGary Federal hacked by Anonymous*, http: //krebsonsecurity.com/2011/02/hbgary-federal-hacked-by-anonymous/, Accessed Dec 2011.

Kumaraguru, P., Rhee, Y., Acquisti, A., Cranor, L. F., Hong, J., & Nunge, E. (2007). Protecting people from phishing: The design and evaluation of an embedded training email system. In *Proceedings of the SIGCHI conference on human factors in computing systems, ser. CHI '07* (pp. 905–914). New York: ACM.

Likarish, P., Dunbar, D., & Hansen, T. E. (2008). Phishguard: A browser plug-in for protection from phishing. In *2 international conference on internet multimedia services architecture and applications, 2008. IMSAA 2008* (pp. 1–6).

Moore, T., & Clayton, R. (2007). Examining the impact of website take-down on phishing. In *eCrime '07: Proceedings of the anti-phishing working groups 2nd annual eCrime researchers summit* (pp. 1–13). New York: ACM.

Prakash, P., Kumar, M., Kompella, R. R., & Gupta, M. (2010). Phishnet: Predictive blacklisting to detect phishing attacks. In *INFOCOM'10: Proceedings of the 29th conference on information communications* (pp. 346–350). Piscataway: IEEE Press.

Rich Skrenta, & Bob Truel. (1998). DMOZ Open Directory Project. http://www.dmoz.org

Salakhutdinov, R. R., & Larochelle, H. (2010). Efficient learning of deep Boltzmann machines. In *Proceedings of the international conference on artificial intelligence and statistics* (Vol. 13).

Schneier, B. (2011). *Lockheed Martin hack linked to RSA's SecurID breach*, http://www.schneier.com/blog/archives/2011/05/lockheed martin.html, Accessed Dec 2011.

Selvaganapathy, S. G., Nivaashini, M., & Natarajan, H. P. (2018). Deep belief network based detection and categorization of malicious URLs. *Information Security Journal: A Global Perspective, 27*(3), 145–161. https://doi.org/10.1080/19393555.2018.1456577.

Sewak, M., Karim, M. R., & Pujari, P. (2018). *Practical convolutional neural network models: Implement advanced deep learning models using Python*. Packt Publishing Ltd. Birmingham, United Kingdom.

Sheng, S., Wardman, B., Warner, G., Cranor, L. F., Hong, J., & Zhang, C. (2009, July). An empirical analysis of phishing blacklists. In *Proceedings of the 6th conference in email and anti-spam, ser. CEAS'09*. Mountain view.

Sheng, S., Holbrook, M., Kumaraguru, P., Cranor, L. F., & Downs, J. (2010). Who falls for phish?: A demographic analysis of phishing susceptibility and effectiveness of interventions. In *Proceedings of the 28 international conference on human factors in computing systems, ser. CHI '10* (pp. 373–382). New York: ACM.

Statistical & Qualitative Data Analysis Software: (2019) About R and RStudio. https://libguides.library.kent.edu/statconsulting/r

Weider, D., Yu, Nargundkar, S., & Tiruthani, N. (July 2008). A phishing vulnerability analysis of web based systems. In *Proceedings of the 13th IEEE symposium on computers and communications (ISCC 2008)* (pp. 326–331). Marrakech: IEEE.

Whittaker, C., Ryner, B., & Nazif, M. (2010). Large-scale automatic classification of phishing pages. In *NDSS '10*.

WHOIS look up., www.whois.net, www.whois.com

Yoshua, B., & Pascal, L. (2007). Greedy layer-wise training of deep networks. In *Advances in neural networks*.

Yue, C., & Wang, H. (2008). Anti-phishing in offense and defense. In *Computer security applications conference, ACSAC 2008. Annual, 8–12 2008* (pp. 345–354).

Efficient Reconfigurable Integrated Cryptosystems for Cybersecurity Protection

Abiy Tadesse Abebe, Yalemzewd Negash Shiferaw, and P. G. V. Suresh Kumar

1 Introduction

Modern cryptosystems include various cryptographic mechanisms designed to defend the cyber-attacks which target different application areas. The major attack types including eavesdropping, data modification, impersonation, repudiation, replay, etc., can be successfully defended by applying symmetric key and asymmetric key algorithms in combination with one or more of the hash functions, Message Authentication Code (MAC) algorithms and digital signature algorithms (Forouzan 2008).

Depending on the application areas and the type of cyber-attacks, cryptosystem implementations require careful selection of suitable algorithms and optimized implementations for successful protection of the attacks. Understanding of the advantages and limitations of the existing algorithms is important before implementing them for securing particular application scenarios. Generally, selection of suitable cryptographic mechanisms for implementation so as to address different cyber-attacks may require careful considerations of the following major questions:

- Is the application area constrained or high performance?
- What are the attack types threatening the specific application area?
- Which cryptographic security services are required to defend the attacks in that specific area?
- Which cryptographic mechanisms can provide the required crypto services?
- Which implementation platform: software based, hardware based, or their combination is suitable for the security of that specific application?

A. T. Abebe (✉) · Y. N. Shiferaw
Addis Ababa Institute of Technology, AAU, Addis Ababa, Ethiopia

P. G. V. S. Kumar
Ambo University, Ambo, Ethiopia

S. K. Shandilya et al. (eds.), *Advances in Cyber Security Analytics and Decision Systems*, EAI/Springer Innovations in Communication and Computing,
https://doi.org/10.1007/978-3-030-19353-9_4

- Does the application involve secure communication between the constrained and high-performance environments? If so, does it require end-to-end security?
- If end-to-end security is required, is it possible to implement a cryptosystem which can fulfil the requirements of both platforms?

The existing cryptographic mechanisms can be effectively utilized to address the above points.

1.1 *Effective Utilization of Existing Cryptographic Mechanisms*

To effectively utilize the advantages of different existing cryptographic mechanisms such as symmetric key and asymmetric key algorithms, hash functions, Message Authentication Code (MAC) generators, and digital signature algorithms, many researchers have proposed various methods to increase the capabilities of the algorithms in terms of performance and security. The authenticated encryption algorithms (Wu and Preneel 2013; McGrew and Viega 2005; Koteshwara and Das 2017) that can provide data confidentiality, data integrity, and data origin authentication services simultaneously using only one algorithm are improvements over symmetric key algorithms. Similarly, signcryption methods (Ullah et al. 2017; Ting et al. 2017) that can provide data confidentiality and digital signature crypto services simultaneously using only one algorithm are improvements over public key algorithms. Despite their advantages, the authenticated encryption and signcryption methods generally share the inherent limitations of symmetric and asymmetric key algorithms. Symmetric key algorithms, though efficient, are limited as they lack key distribution mechanism, which requires sharing of a secret key among communicating parties before starting secret information exchange. The asymmetric key algorithms have avoided the requirement of sharing of the secret key by providing a key pair, one for encryption and another for decryption, but their performance is generally considered slower because of the intensive mathematical operations needed to accomplish encryption and decryption processes.

To overcome these limitations, various researchers have proposed hybrid cryptosystems (Gutub and Khan 2013; Kapur and Khatri 2015; Alkady et al. 2013) and Integrated Encryption Schemes (Abdalla et al. 2001; Martínez et al. 2015). The main goal of these methods is to effectively utilize the advantages of different cryptographic mechanisms, particularly, the symmetric key and the asymmetric key algorithms so that the combined cryptosystem can use the symmetric key algorithm for large amount of data encryption and decryption while using the asymmetric key algorithm for key distribution. As a result, cryptosystems as efficient as symmetric key algorithms and as secured as asymmetric key algorithms have been developed. Moreover, more number of crypto services have been provided by integration of hash functions, MAC algorithms, and digital signature algorithms with the hybrid systems or integrated encryption schemes for strong security.

1.2 Consideration of the Application Areas

Cryptosystem implementations for high-performance applications and for constrained environments must be different. Generally, high-performance platforms have enough resources with high computing power and, therefore, require cryptosystems with high throughput and strong security. The constrained platforms, on the other hand, have limited resources and low computing power, requiring lightweight cryptosystems. As a result, a cryptosystem designed for high-performance platforms may make constrained platforms inefficient if directly used, and lightweight cryptosystems designed for constrained platforms may not meet the high throughput and strong security requirements of the high-performance platforms. Therefore, the application areas have to be considered for successful cryptosystem implementations (Suárez-Albela et al. 2019).

1.3 Attack Types and the Required Cryptographic Security Services

Though there are various cyber-attacks, they can be categorized within the major attack types such as eavesdropping, data modification, impersonation, replay, and repudiation attacks (Forouzan 2008). As the attack surface increases, more crypto services may be required to address them. However, the crypto services needed for securing a particular application depends on the security requirement. Some applications may require data integrity check only. Some may require authentication only. Some other applications may require data confidentiality service only or data confidentiality with data integrity and authentication. Other applications may also require confidentiality, integrity, authentication, and nonrepudiation services. Therefore, the application's specific security requirement has to be considered before cryptosystem implementation.

1.4 End-to-End Security

If the application scenario requires secure communication between constrained and high-performance platforms, the cryptosystem is needed to fulfil the specific requirements of each platform. The two platforms are different in terms of available resources as well as performance and security requirements. It is a challenging task to implement a lightweight cryptosystem which meets the requirements of constrained devices while simultaneously fulfilling the high-throughput, low-latency, and strong security requirements of the high-performance platforms and vice versa. In addition, the secret information exchanges between constrained and high-performance platforms should not reveal secret data at intermediate steps demanding end-to-end security, which is also a challenging task to implement using the same cryptosystem (Raza et al. 2017; Patel and Wang 2010; Moosavi et al. 2016).

1.5 Implementation Platforms

Based on the available resources, performance, and security requirements of a particular application area, software-based or hardware-based implementations of cryptosystems can be preferred. The implementation flexibility and ease of modification make software-based implementations preferable. However, compared to their hardware counterparts, they are slower especially for high-speed applications. Hardware-based implementations of cryptosystems are useful, in one hand, for providing high-speed performance, and, on the other hand, for providing physical security. Depending on the application scenarios, FPGA-based implementations are more preferable because of several reasons. One of the reasons is that they can be reconfigured based on the contemporary attack situations. The other reason is that they can be as flexible as software for implementations and as high speed as hardware in terms of performance (Dube 2008; Wanderley et al. 2011).

In this chapter, efficient FPGA implementation of integrated cryptosystems are proposed for high-performance platforms, constrained devices, and secure information exchanges between constrained and high-performance platforms, targeting the security of healthcare IoT application scenario.

The rest of the chapter is organized as follows: Section 2 describes related works. Section 3 briefly describes the selected algorithms. Section 4 presents cryptosystems proposed for high-performance platforms. Section 5 explains the cryptosystem proposed for a healthcare IoT which consists of constrained and high-performance platforms. Finally, Section 6 concludes the chapter.

2 Related Works

Different researchers have proposed various hybrid cryptosystems and Integrated Encryption Schemes to improve the limitations of symmetric and asymmetric key algorithms. In such cryptosystems, various algorithms have been incorporated in one system for producing more crypto services to get strong security. For example, Diffie-Hellman Integrated Encryption Scheme (DHIES) and Elliptic Curve Integrated Encryption Scheme (ECIES) are standardized methods integrating different cryptographic mechanisms including a key agreement protocol, a Key Derivative Function (KDF), an encryption algorithm, a hash function, and a Message Authentication Code algorithm (Abdalla et al. 2001; Martínez et al. 2015). In some hybrid cryptosystems, digital signature algorithms have been also included (Kapur and Khatri 2015; Alkady et al. 2013). However, most of the existing hybrid cryptosystems (Gutub and Khan 2013; Kapur and Khatri 2015; Alkady et al. 2013) and the Integrated Encryption Schemes (Abdalla et al. 2001; Martínez et al. 2015) have been designed targeting high-performance platforms. They will be ineffective if they are implemented directly for securing constrained environments such as wireless body area networks and constrained IoT devices since such devices do not have enough resources and computing capabilities as well as sufficient energy to run these algorithms (Suárez-Albela et al. 2019). Moreover, these cryptosystems have combined four or more different algorithms to provide more cryptographic security services including data confidentiality, data integrity, authentication, non-repudiation, etc., but each component algorithm requires a separate key for security purpose. The key management, key storage, and compatibility issues as well as overall storage space requirement for the cryptosystem need critical considerations. In addition, DHIES and ECIES have been designed based on a unilateral key confirmation approach (Barker et al. 2018; Tadesse Abebe et al. 2019) such that verification of the authenticity of only one end of the communicating parties is possible, but the corresponding end at the other side communicates without authenticating the corresponding end to exchange the secret key and secret information. But, exchanging secret information without verifying the authenticity of the entity communicating with at the other end can make the communication vulnerable to security risks due to man-in-the-middle attacks (Barker et al. 2018; Tadesse Abebe et al. 2019).

On the other hand, most of the existing lightweight cryptosystems that have been proposed for securing constrained platforms are based on symmetric key techniques (Biryukov and Perrin 2018; Okello et al. 2017) and lack key distribution methods. They will not also satisfy the requirements of the high-performance applications since such applications require strong security, high throughput, and need more number of cryptographic services as the attack surface is more.

3 The Selected Algorithms

In this section, we briefly describe the reason why the crypto algorithms implemented in this study and included in this chapter are selected.

3.1 Cryptographic Algorithms Selected for High-Performance Platforms' Security

High-performance applications require high-speed cryptosystems to provide high throughput and low latency along with major cryptographic services for strong security.

For the security of high-performance platforms, AEGIS-128 Authenticated Encryption with Associated Data (AEAD) algorithm (Wu and Preneel 2013), Diffie-Hellman (DH) key exchange protocol (Diffie and Hellman 1976), and SHA-256 hash function (Federal Information Processing Standards 2015) are selected. AEGIS-128 is an AES (FIPS Publication 197 2001) based algorithm and has efficient performance which can be optimized to suite the high-throughput and low-latency requirements of high-performance platforms. The DH algorithm is selected for key exchange instead of using key transport method due to its simplicity for implementation. SHA-256 algorithm is selected to work as a Key Derivation Function (KDF) to generate 256 key material suitable for the AEAD algorithm from the output computed by the DH algorithm. Readers can refer to Refs. (Wu and Preneel 2013; Diffie and Hellman 1976; Federal Information Processing Standards 2015; FIPS Publication 197 2001; Krawczyk and Eronen 2010) for detailed constructions of the selected algorithms.

3.2 Cryptographic Algorithms Selected for Constrained Devices' Security

As the constrained devices have limited resources and lower computation capability, they cannot not accommodate the algorithms that are suitable for the high-performance platforms. They require lightweight cryptosystems with smaller area, low power consumption, and good performance.

For the security of constrained platforms, ASCON-128 lightweight AEAD algorithm (Dobraunig et al. 2016) is selected. ASCON is a flexible algorithm with ASCON-128 and ASCON-128a varieties providing 64-bit and 128-bit blocks of data processing capability, respectively. Optimized implementation of ASCON is possible for area and speed tradeoffs. Readers can refer to Ref. (Dobraunig et al. 2016) for detailed constructions of the selected algorithm.

4 Implementation of Integrated Cryptosystems for High-Performance Platforms

4.1 Description of the Method

A cryptosystem using authenticated encryption algorithm and a key exchange with bilateral key confirmation method is implemented by integrating the Diffie-Hellman (DH) key agreement protocol and AEGIS-128-authenticated encryption algorithm as shown in Figs. 1, 2 and 3. The SHA-256 algorithm is used as a KDF. The details of the AEGIS-128 algorithm, DH key agreement protocol, SHA-256, and KDF are available in the literature. In this chapter, we present only our proposed ideas, the structure of the proposed methods, the implementation approaches, and the implementation outcomes. The proposed method is explained based on three major steps including key exchange, bilateral key confirmation, and authenticated encryption and authenticated decryption processes.

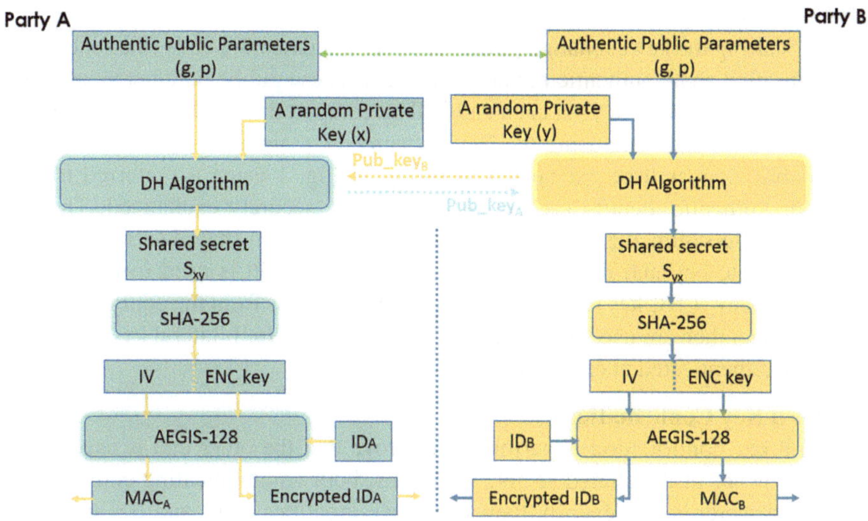

Fig. 1 Key agreement, secret key and *IV*generation, and authenticated encryption of IDs

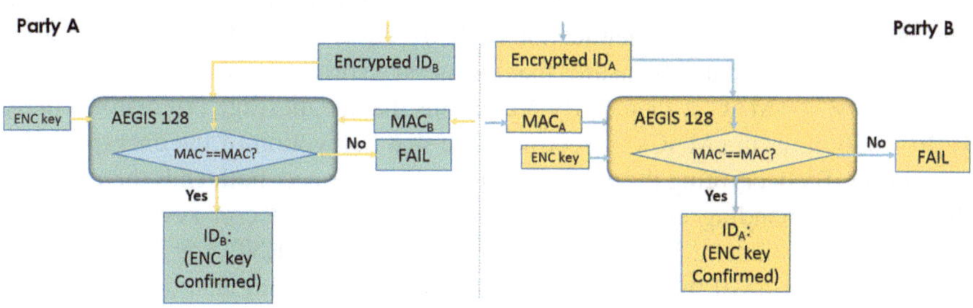

Fig. 2 Bilateral key confirmation between high-performance platforms

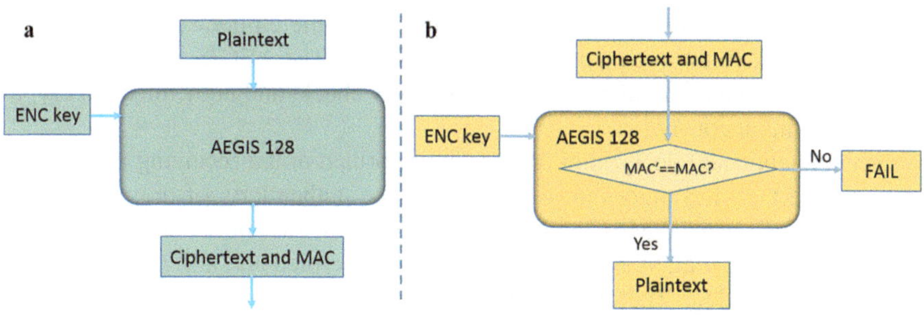

Fig. 3 Secure message exchange: (**a**) Authenticated encryption. (**b**) Authenticated decryption

Key Exchange

Before starting the key exchange process, both communicating ends (the sender and recipient) first share authentic public parameters. These public parameters and randomly selected private keys at each end will be computed to produce the respective public keys for each end using the DH key agreement algorithm. Then, the generated public keys will be exchanged as shown in Fig. 1 using a preferred trusted method. Authentic identity credentials can also be securely exchanged. The key agreement algorithm at each end then computes a common shared secret using the owned private key and the received public key. SHA-256 is used to produce an encryption key (ENC key) and *IV* with appropriate lengths from the common shared secret, which then are used by the AEGIS-128 algorithm for bilateral key confirmation and authenticated encryption.

Bilateral Key Confirmation

Bilateral key confirmation is performed to prove that the other end is exactly the claimed owner of the shared secret. To do this, both communicating ends use the generated secret key to encrypt their respective identities (ID_A, and ID_B)using the AEGIS-128 algorithm and exchange the encrypted data and the corresponding Message Authenticated Codes (MAC_A and MAC_B) to the other end as shown in Fig. 1. The AEGIS 128 algorithm at each end then compares the received MAC and the calculated MAC' for authentication of the IDs as shown in Fig. 2. For example, if $MAC_A = MAC'_A$, party B ensures that the message originator is party A; and similarly, if $MAC_B = MAC'_B$, then, party A confirms that the message originator is party B. It is only after the MACs are verified true that decryption of the encrypted IDs will follow. By decrypting the encrypted ID of the other end, both ends then confirm that the corresponding end is the claimed owner of the secret key, meeting the requirement of bilateral key confirmation. The computed ENC key is not directly used for encryption of the sensitive data without first verifying that the other end is the intended owner of that ephemeral key.

Authenticated Encryption and Authenticated Decryption

After bilateral key confirmation, the sender encrypts the secure message using the secret key and the AEGIS-128 algorithm and sends the ciphertext and the corresponding MAC to the recipient as shown in Fig. 3a. The AEGIS-128 algorithm is fast and can provide confidentiality, data integrity check, and authentication crypto services simultaneously.

The authenticated decryption process is performed at the receiving end. In this case, the AEGIS-128 algorithm first validates the authenticity of the message by comparing the received MAC and the calculated MAC' as shown in Fig. 3b. If the MAC values are equal, then the message will be decrypted and utilized. Otherwise, it will be discarded. Therefore, using the proposed method, authenticated key exchange along with bilateral key confirmation and authenticated encryption/decryption is possible.

4.2 Implementation Approaches

In this work, an integrated scheme combining the Diffie-Hellman key-agreement protocol, SHA-256 KDF, and AEGIS-128 AEAD algorithm is implemented on Xilinx Virtex 5, Virtex 7, and Virtex II FPGA devices and synthesized for comparison with the existing works using Xilinx ISE 14.5, Vivado Design Suite 2017.2, and Xilinx ISE 10.1, respectively. VHDL is used as a hardware description language. It is assumed that the public parameters are authentic and shared between the two communicating ends before starting secure communication. Also, it is assumed that the two ends have exchanged the generated public keys at each end using a chosen trusted method. SHA-256 is used to produce 256-bit length value from the computed shared secret at both ends generating 128-bit ***IV*** and 128-bit secret key, which are suitable for the AEGIS-128 algorithm. A pipelined AES (Satoh et al. 2007) has been implemented for AEGIS-128. In pipelining architecture, registers have been placed at each step/round to construct the pipeline as shown in Fig. 4. The depth of the pipeline, ***K***, determines how many data blocks can be processed concurrently. The architecture is fully pipelined when ***K*** equals the total number of rounds. The area and the latency of the pipelined architecture are proportional to ***K***. Pipelining can increase the encryption speed by processing multiple blocks of data simultaneously. The five AES rounds in AEGIS-128 algorithm are pipelined to concurrently process and update the state of AEGIS-128 to provide high throughput in a clock cycle. Loop unrolling on iterative operations and pipelining techniques are applied for concurrent or parallel operations of different constructs to increase the throughput with some area cost.

For implementation of the Diffie-Hellman key agreement algorithm, a Montgomery multiplier (Montgomery 1985) has been used to perform the modular multiplication which is used to speed up the process. The public parameters selected for implementation of the DH protocol are $p = 991$ bits and $q = 503$ bits. Optimization of SHA-256 is performed using loop unrolling and pipelining along with a hardware-sharing technique.

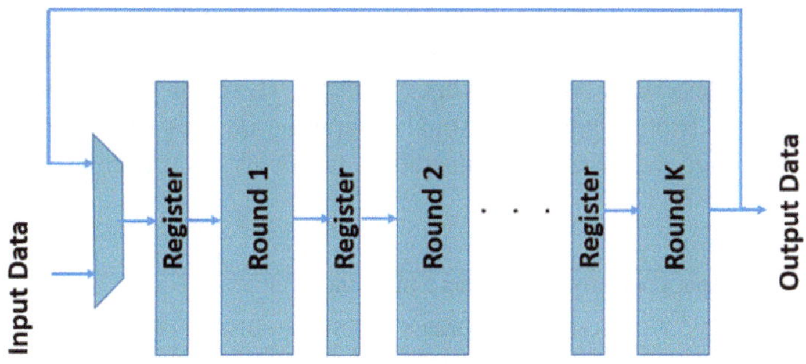

Fig. 4 Pipeline architecture (Tadesse Abebe et al. 2019)

4.3 Results

The implementation results and the performance comparisons with existing works are shown in Tables 1, 2, and 3. Table 1 shows the result comparison in terms of utilization of FPGA resources, maximum frequency, and the achieved throughput (TP) for the Virtex 5 device. As shown in Table 1, the optimization method used by (Abdellatif et al. 2016) is LUT based, and the present work used pipelining method. The implementation target of our work is high throughput. Therefore, for the Virtex 5 device, the present work achieved a throughput of 44.6 Gbps with an area cost of 5586 slices and 6 BRAMs. For the Virtex 7 device, compared to the result presented by (Katsaiti and Sklavos 2018), the present work achieved smaller area and better throughput for the same pipelined optimization approach as shown in Table 2. Comparison of the consumed FPGA resources for the whole implemented integrated cryptosystem of the present work with FPGA-based ECIES implementation results reported by (Sandoval and Uribe 2005) on Virtex II device is shown in Table 3. In this case, the outcome of the present work shows smaller space utilization (slices and BRAMs) compared to outcome reported by (Sandoval and Uribe 2005).

Unlike the existing works presented in Tables 1 and 2 where the implementations were restricted to performance enhancement and utilization of optimized FPGA resources leaving the lack of key distribution method untouched, the contributions of this work include authenticated encryption and authenticated key distribution with bilateral key confirmation offering strong security with more crypto security services in addition to performance enhancement and reasonable FPGA resources utilization.

5 Healthcare IoT Comprising Constrained and High-Performance Platforms

In this section, we consider the security case of healthcare Internet of Things (IoTs) that incorporates both constrained and high-performance platforms. This e-health system can consist of wearables or biosensor nodes deployed around a human body forming a Wireless Body Area Network (WBAN), which can then connect the sensor nodes to each other and to personal computing devices such as smartphones using short-range wireless communication protocols to transmit physiological

Table 1 Performance comparison for Virtex 5 platform

Author	Target device	Design	Slices	BRAM	Freq. (MHz)	TP (Gbps)
This work	Virtex 5	Pipelined	5586	6	348.7	44.6
Abdellatif et al. (2016)	Virtex 5	LUT based	1391	0	156.5	20.03

Table 2 Performance comparison for Virtex 7 platform

Author	Target device	Design	Slices	TP (Mbps)
This work	Virtex 7	Pipelined	9306	89354
Katsaiti and Sklavos (2018)	Virtex 7	Pipelined	10610	88564

Table 3 Performance comparison for Virtex II platform

Author	Target device	Design	Slices	BRAMs
This work	Virtex II	Hybrid of AEGIS + DH + SHA-256	14572	4
Sandoval and Uribe (2005)	Virtex II	ECIES	21194	20

parameters gathered from the patient's body. The personal computing devices can then communicate with remote central servers through heterogeneous communication networks to transmit health-related critical data for remote analysis, processing, and real-time monitoring. This application scenario can be conceptually represented as shown in Fig. 5. The system consists of WBAN, intermediate devices (smartphone/laptop), and high-speed servers that can communicate wirelessly. The WBAN, with its biosensors and wearables, is considered a constrained platform while the other parts are considered high-performance platforms. The intermediate devices may not have comparable performance with the high-speed servers but compared to the constrained sensor nodes, they are considered to have enough computing power and resources to run integrated cryptosystems.

The Internet or Cloud IoTs can be used to establish healthcare cloud services for remote health data access (Cirani et al. 2019). Intermediate devices such as smartphones, Personal Digital Assistants (PDAs), laptops, or other PCs can wirelessly communicate to the WBAN using Bluetooth Low Energy (BLE), ZigBee, or other suitable standardized wireless communication protocols to receive the data transmitted by the Intra-BAN. These intermediate devices can also communicate to the Internet or the Cloud IoT so that applications in high-speed servers can process, analyze, or store the data so that medical advisors can make real-time decisions remotely based on the processed data using the cloud services. Based on the decision, the patient can get treatment, advice, prescriptions, or any other cautions by the medical advisor and local care givers. Families can also follow the patient's health conditions and the healthcare process remotely. In case of emergency, fast emergency aid can be provided based on the emergency signals or alarms generated by the system (Raza et al. 2017; Moosavi et al. 2016). In Fig. 5, only one patient is shown for demonstration purpose but the healthcare system can commonly incorporate more number of patients in a similar fashion.

For the case of IoT-based remote healthcare, these two different platforms (constrained and high performance) have to securely communicate to each other in order to exchange health critical data by means of end-to-end security.

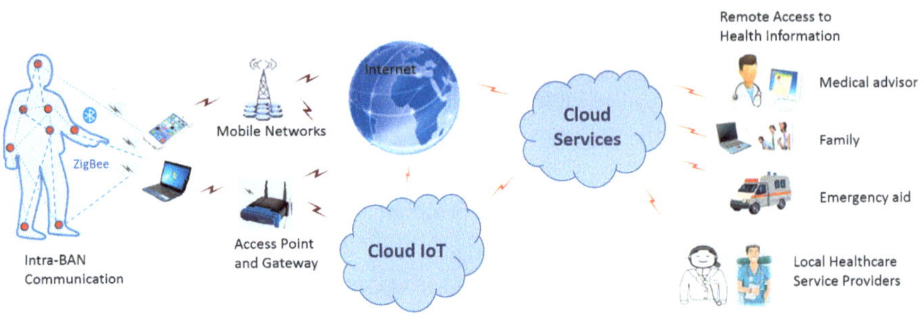

Fig. 5 Conceptual representation of IoT-based remote health monitoring

5.1 *The Proposed Method*

The proposed method is based on the remote healthcare application scenario represented in Fig. 5. It is implemented considering the three important steps of the whole system including the constrained platforms, intermediate devices (smartphone/laptop/PDA), and the central server. The same authenticated encryption algorithm is used for secure communication between the two (constrained and high-performance) platforms. But, for the high-performance platforms, the algorithm is modified and enhanced to achieve the specific requirements. A key agreement protocol with bilateral key conformation is integrated with the authenticated encryption algorithm for the high-performance platforms' security. In case of the constrained devices, it is assumed that secret keys can be sent periodically as required from the high-performance platform side depending on the risk situation as they have not enough resources and computing power to run an integrated key exchange scheme. However, the authenticated encryption algorithm of the constrained devices can verify the authenticity and check the integrity of the received data whenever they receive secret keys or any other secret data.

Constrained Platform
The lightweight ASCON-128 algorithm (Dobraunig et al. 2016) is implemented on FPGA and optimized for smaller area and good performance for the constrained devices. It processes smaller amount of data (64 bits). Figure 6 shows the general structure of authenticated encryption (Fig. 6a) and authenticated decryption (Fig. 6b) of the architecture.

The optimized lightweight ASCON-128 algorithm provides the required security services including data confidentiality, data integrity, and authentication simultaneously. This helps to send secret information with verification data to the high-performance devices and also to verify data integrity, validate authentication, and decrypt encrypted data whenever the constrained devices receive messages and secret keys from the high-performance platforms.

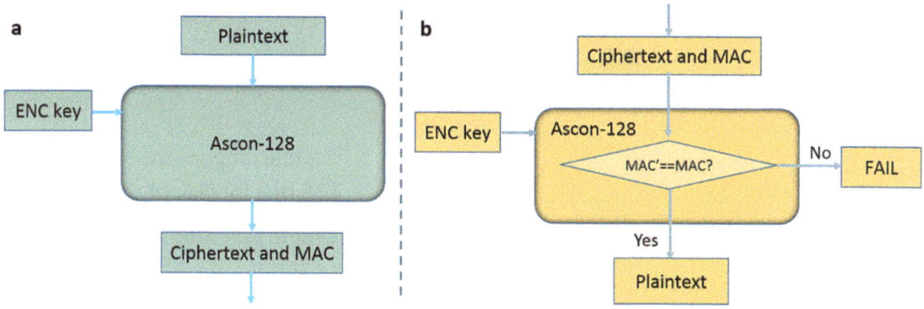

Fig. 6 Lightweight ASCON algorithm for constrained platforms: (**a**) Authenticated encryption (**b**) Authenticated decryption

High-Performance Platforms

For high-performance platforms, ASCON-128 and ASCON-128a algorithms are integrated so that the hybrid system can provide the functions of both algorithms in parallel for processing 64-bit and 128-bit blocks of data, respectively. The integrated ASCON algorithm can process the 64-bit and 128-bit blocks of data in parallel to speed up the overall operations as it can communicate with the constrained platform using 64-bit blocks of data processing (ASCON-128) and also between the high-performance platforms using 128-bit blocks of data processing (ASCON128a).

For key distribution, the Diffie-Hellman (DH) key agreement algorithm (Diffie and Hellman 1976) and SHA-256 Key Derivative Function (KDF) (Federal Information Processing Standards 2015; Krawczyk and Eronen 2010) are used with the bilateral key confirmation method integrated with the ASCON algorithm to provide the required security services. ASCON can satisfy the requirements of the constrained platforms and, with modification, enhancement, and optimization of it for high throughput, it can also be made suitable for high-performance applications so that the constrained devices and the high-performance platforms can securely communicate.

Intermediate Devices

When receiving the encrypted data with the corresponding Message Authentication Code (MAC) from a constrained device which uses the ASCON-128 AEAD algorithm, the intermediate platform performs the data aggregation process first, then, the received 64-bit ciphertext blocks are rearranged to 128-bit blocks, and authenticated encryption is performed again using the ASCON-128a algorithm for strong security and enhanced performance as well as to contribute to the end-to-end security of the system as shown in Fig. 7. This has been performed since the lengths of the plaintext and the ciphertext are equal for the ASCON AEAD algorithm (Dobraunig et al. 2016). The ephemeral secret key for this process is obtained using the key exchange mechanism shown by Figs. 9 and 10. As shown in Fig. 7, the *ENC key* is the secret key which is the same as the secret key used by the constrained device and the *ENC key1* is the ephemeral key used by the high-performance platforms.

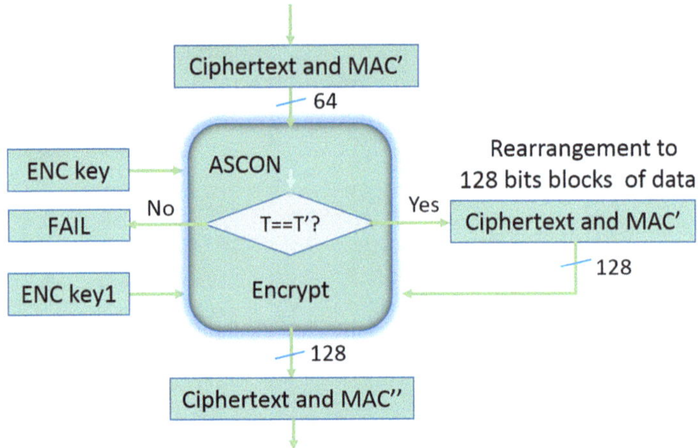

Fig. 7 Data rearrangement, and authenticated encryption at the intermediate step

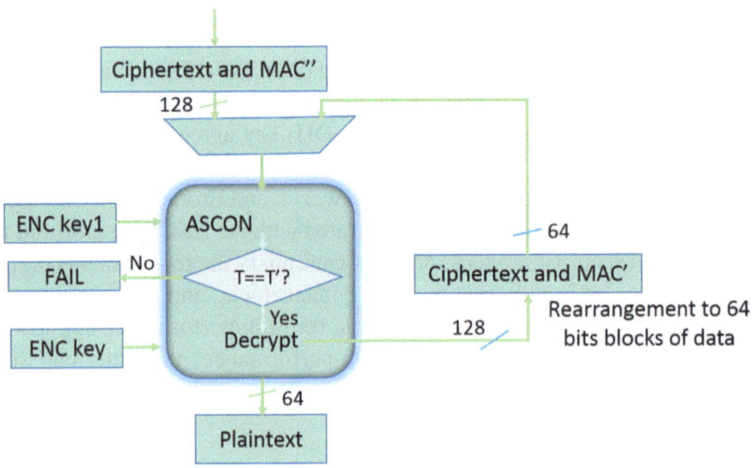

Fig. 8 Double-step verifications and authenticated decryptions at the central server

Server Side

The central server will perform two phases of verification and decryption processes in order to recover the original data. First it performs the verification of the MAC created by the intermediate device. If the received MAC and the calculated MAC are equal, then it performs decryption of the encrypted data (128-bit ciphertext blocks) sent by the intermediate device. Otherwise (if the MAC comparison fails), the data will be discarded as shown in Fig. 8. If the verification is valid, then the decrypted ciphertext is rearranged into the original 64-bit blocks of data as was previously sent by the constrained device. The MAC comparison is performed again for this data. If the received MAC and the calculated MAC are equal, then the final decryption process will produce the original plaintext; otherwise the entire data will be discarded.

Key Exchange

The key exchange and the bilateral key confirmation methods are similar to the methods explained in Sect. 4.1 with the exception that in this case the hybrid of the ASCON-128 and ASCON-128a AEAD algorithms is used instead of the AEGIS-128 algorithm. Therefore, for authenticated key distribution with bilateral key confirmation, a cryptosystem that integrates DH key exchange protocol, SHA-256 Key Derivative Function, and hybrid ASCON authenticated encryption algorithm is used as shown in Figs. 9 and 10. The DH algorithm performs key exchange while the KDF produces the required size of 128-bit ***IV*** and 128-bit secret key for the ASCON algorithm. Then, the ASCON algorithm performs authenticated encryption of the identification of each end so that each end authenticates each other using the newly generated secret key before using the secret key for encryption of any secret data. Then, they exchange the encrypted IDs and the MAC values as shown in Fig. 9. The IDs of the respective ends that have been already exchanged securely are encrypted with time stamps and any other important identification credentials useful for authentication and ensuring the data freshness.

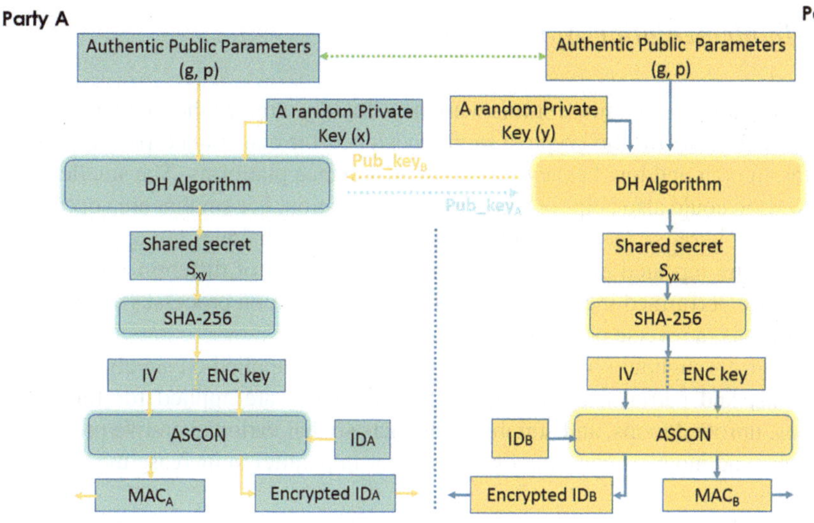

Fig. 9 Key exchange between high-performance platforms

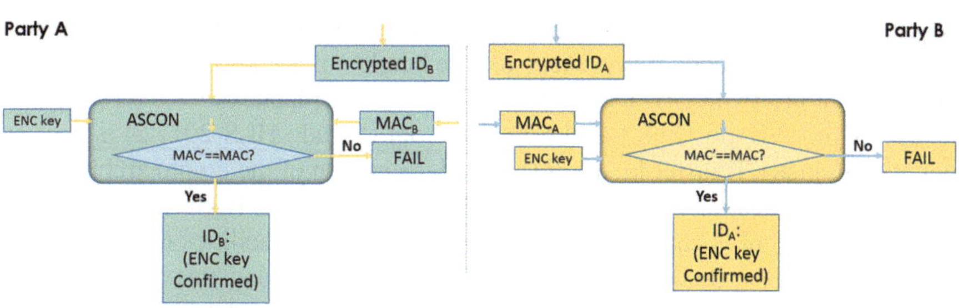

Fig. 10 Bilateral key confirmation technique

Bilateral Key Confirmation

Upon receiving of the encrypted ID and the MAC of the other end, each end will then perform comparison of the received MAC and the calculated MAC first to check data integrity and authentication. Only after this verification step is true that the decrypted IDs will be utilized. Otherwise, the data will be rejected. If the verification step is valid and the encrypted IDs are decrypted with the ephemeral secret key, both ends will confirm that they are communicating with a legitimate body based on bilateral key confirmation approach as shown in Fig. 10 and then start a trustworthy information exchange.

The security of the proposed integrated cryptosystem can be expressed in terms of data confidentiality, data integrity, authentication, exchange of ephemeral keys with bilateral key confirmation method, end-to-end security without revealing secret data at intermediate steps, and double-phase protection of data in the high-performance environment. Therefore, it can detect and prevent eavesdropping, data modification, impersonation, and replay attacks.

5.2 Implementation Approaches

The optimization techniques for this work have mainly targeted the smaller area and good performance for the cryptosystem envisioned for constrained platforms, and high throughput and low latency for high-performance platforms. The smaller area optimization could also help for low power optimization. For smaller area optimization, mainly, iterative methods and hardware-sharing techniques are applied wherever the same repeated processes and similar operations of different functions or modules are performed including encryption and decryption processes where similar operations in reverse order are performed.

For high-throughput optimization, loop unrolling is applied on suitable iterative operations, and partial and full pipelining techniques are applied for functions, modules, unrolled loops, and suitable constructs so that various iterative processes, functions, and modules can run concurrently or in parallel to increase the throughput with area cost.

A hybrid of ASCON-128 and ASCON-128a is implemented for the high-performance platform and optimized using hardware-sharing technique so that it can switch from 64-bit data processing to 128-bit data processing and vice versa as required, since these ASCON algorithm varieties have only small differences such as the size of data blocks they process and the number of rounds used for intermediate permutations to process the associated data and the plaintext.

The Xilinx Vivado High Level Synthesis tool (Vivado HLS 2017.2) is used to generate Register Transfer Level (RTL) implementation of the crypto architectures from C++ specifications for synthesis into FPGA. The Vivado High Level Synthesis (Vivado HLS) tool is equipped with various optimization facilities (Xilinx Vivado Design Suite User Guide High-Level Synthesis 2018). Using this HLS tool, the required optimization targets including throughput, latency, and power are explored and applied. This approach is used to utilize the optimization facilities of the Vivado High Level Synthesis tool for optimization of the proposed method in addition to our implementation techniques applied in realizing the proposed technique. Each module was implemented separately and the functionality of each was tested independently. Test benches were developed for each module and then C-simulation, C-synthesis, and C/RTL co-simulation were performed. Then, the resulting VHDL code of the RTL design is applied to the Xilinx ISE Design Suite 14.5 and synthesized into the Xilinx Spartan 6 device for the purpose of comparison with the same existing implementations as most of recent existing implementations have used the Spartan-6 device for their implementations (Yalla and Kaps 2017; Diehl et al. 2018; Farahmand et al. 2018).

For the implementation of the Diffie-Hellman key agreement algorithm, the Montgomery multiplier (Montgomery 1985) has been used to perform the modular multiplication so as to speed up the computation process. The public parameters selected for implementation of the DH protocol are $p = 991$ bits and $q = 503$ bits. Optimization of SHA-256 is performed using loop unrolling and pipelining along with the hardware-sharing technique.

5.3 Results

The implementation outcomes are described with reference to the optimizations performed for the corresponding intended specific platform. Table 4 shows the performances achieved and resources utilized for the cryptosystem implemented for the constrained device (ASCON-128) and the high-performance platforms (hybrid ASCON) denoted as H_ASCON. The results show that the cryptosystem implemented for the constrained device produced 252.86 Mbps throughput (TP) and an area of 658 LUTs. The cryptosystem implementation for high-performance platforms consumed more hardware resources (4072 LUTs) than the cryptosystem implemented for the constrained device. However, concerning the throughput, 6874.5 Mbps is achieved for the cryptosystem implemented for the high-performance platform. The results are also separately presented as lightweight and high performance for the constrained and high-performance platforms, respectively, as shown in Table 5 and Table 6 for comparison with the existing similar works as they have used this approach to present their implementation outcomes. Table 5 shows the implementation results for lightweight implementations and Table 6 shows the implementation outcomes for high performance. Compared to the existing works, the hardware resources utilized by the present work is smaller (658 LUTs) compared to 684 LUTs and 2048 LUTs achieved by (Yalla and Kaps 2017; Diehl et al. 2018). The throughput of 252.86 Mbps achieved by the present work is better than the 60.1 Mbps and 119.16 Mbps for ASCON-128 and ASCON-128a of (Yalla and Kaps 2017), respectively, and closer to the 255.4 Mbps of (Diehl et al. 2018) which consumed more than three times resources than the present work.

Table 4 Implementation results for different platforms

Platform	FPGA Device	Implementation	Area LUT	Freq. (MHz)	TP (Mbps)	TP/ Area
Constrained	Spartan-6	ASCON-128	658	194.32	252.86	0.384
High performance	Spartan-6	H_ASCON	4072	483.35	6874.5	1.688

Table 5 Comparison with the existing lightweight implementations

Authors	Device	Implementation	Area LUT	Freq. (MHz)	TP (Mbps)	TP/ Area
Yalla and Kaps (2017)	Spartan-6 FPGA	ASCON 128 ASCON-128a	684 684	216.0 216.0	60.1 119.16	0.26 0.52
Diehl et al. (2018)	Spartan 6 FPGA	ASCON	2048	195.5	255.4	0.125
This work	Spartan-6 FPGA	ASCON-128	658	194.32	252.86	0.384

Table 6 Comparison with the existing high-performance implementations

Authors	Device	Implementation	Area LUT	Freq. (MHz)	TP (Mbps)	TP/ Area
Farahmand et al. (2018)	Spartan-6 FPGA	ASCON-128 ASCON-128a	1402 1712	208.5 202.8	1906.3 2884.3	1.360 1.684
This work	Spartan-6 FPGA	H_ASCON	4072	483.35	6874.5	1.688

Concerning the cryptosystem implementation for the high-performance platforms, Table 6 shows the highest throughput of 6874.5 Mbps compared to the existing 1906.3 Mbps and 2884.3 Mbps of (Farahmand et al. 2018) for ASCON-128 and ASCON-128a, respectively. Regarding the hardware resource utilization of implementation for high performance, the present work used 4072 LUTs which is more than the existing achievements: 1402 LUTs and 1712 LUTs as shown in Refs. (Farahmand et al. 2018) for ASCON-128 and ASCON-128a implementations, respectively, as expected, since our optimization target is high throughput for high-performance case with some hardware cost.

6 Conclusions

Differences in performance, security, and available resources must be considered before implementing cryptosystems for different application environments. The selection of suitable crypto algorithms and optimized implementations based on the specific requirements of each platform is also important. Optimization of the cryptosystem for high throughput and low latency are the requirements for high-performance platforms, whereas constrained devices require small area and low power consumption. IoT-based healthcare applications that incorporate constrained and high-performance platforms for secure wireless information exchange require end-to-end security to protect the life critical data and privacy of patients. In this chapter, an integrated cryptosystem with authenticated encryption and key exchange with the bilateral key confirmation approach is implemented for secure information exchange between high-performance applications using the AEGIS-128 AEAD algorithm, DH key agreement protocol, and SHA-256 hash function on FPGA. For secure information exchange between constrained and high-performance applications, the case of healthcare IoT is considered. In this case, ASCON-128 is used for constrained devices and an integrated scheme incorporating the hybrid of ASCON-128 and ASCON-128a algorithms and the DH key exchange algorithm with the bilateral key confirmation method is implemented for the high-performance platform side of the system. In all the cases, authenticated encryption and authenticated key exchange are possible with end-to-end security features. Compared to the existing hybrid cryptosystems that provided only one-sided key confirmation, bilateral key confirmation allows both communicating ends to create a trustworthy communication. Integrating fewer number of algorithms helped to reduce key management, key storage, and overall hardware resource requirements while providing the major crypto services compared to the existing integrated encryption schemes. The implementation outcomes show that the proposed hybrid cryptosystems have consumed reasonable amounts of FPGA resources with better throughput achievements, which we will further improve by applying more optimization techniques.

References

Abdalla, M., Bellare, M., & Rogaway, P. (2001). The oracle Diffie-Hellman assumptions and an analysis of DHIES. Topics in cryptology - CT-RSA. In *Proceedings* (pp. 143–158). Berlin: Springer.

Abdellatif, K. M., Chotin-Avot, R., & Mehrez, H. (2016). *AES-GCM and AEGIS: Efficient and high speed hardware implementations*. New York: Springer Science+Business Media.

Alkady, Y., Habib, M. I., & Rizk, R. Y. (2013). A new security protocol using hybrid cryptography algorithms. *IEEE international computer engineering conference (ICENCO)*, pp. 109–115.

Barker, E., Chen, L., Roginsky, A., Vassilev, A., & Davis, R. (2018, April). Recommendation for pair-wise key-establishment schemes using discrete logarithm cryptography. NIST Special Publication 800-56A Revision 3.

Biryukov, A., & Perrin, L. (2018, Jan). State of the art in lightweight symmetric cryptography. *Cryptology*.

Cirani, S., Ferrari, G., Picone, M., & Veltri, L. (2019). *Internet of Things: Architectures, Protocols and Standards*. John Wiley & Sons. Hoboken, NJ.

Diehl, W., Abdulgadir, A., Farahmand, F., Kaps, J. -P., & Gaj, K. (2018). *Comparison of cost of protection against differential power analysis of selected authenticated ciphers*. IEEE international symposium on hardware oriented security and trust (HOST).

Diffie, W., & Hellman, M. (1976). New directions in cryptography. *IEEE Transactions on Information Theory, 22*(6), 644–654.

Dobraunig, C., Eichlseder, M., Mendel, F., & Schläffer, M. (2016). Ascon v1.2. Submission to the CAESAR competition: 4, 20, https://competitions.cr.yp.to/round3/asconv12.pdf.

Dube, R. R. (2008). *Hardware-based computer security techniques to defeat hackers from biometrics to quantum cryptography*. John Wiley & Sons, Inc., Hoboken, New Jersey.

Farahmand, F., Diehl, W., Abdulgadir, A., Kaps, J. -P., & Gaj, K. (2018). *Improved lightweight implementations of CAESAR authenticated ciphers*. IEEE 26th annual international symposium on field-programmable custom computing machines (FCCM).

Federal Information Processing Standards (FIPS) Publication 180–4. (2015). Secure Hash Standard (SHS), vol. 4.

FIPS Publication 197, the Advanced Encryption Standard (AES), *U.S. DoC/NIST*, (2001, Nov).

Forouzan, B. A. (2008). *Cryptography and network security* (pp. 1–10). Tata McGraw-Hill Publishing Companies, Inc. 7 West Patel Nagar, New Delhi.

Gutub, A. A., & Khan, F. A. (2013). *Hybrid crypto hardware utilizing symmetric-key & public-key cryptosystems*. IEEE international conference on advanced computer science applications and technologies (ACSAT), pp. 116–121.

Kapur, R. K., & Khatri, S. K. (2015). Secure data transfer in MANET using symmetric and asymmetric cryptography. IEEE international conference on reliability, infocom technologies and optimization (ICRITO) (trends and future directions), pp. 1–5.

Katsaiti, M., & Sklavos, N. (2018). Implementation efficiency and alternations, on CAESAR Finalists: AEGIS Approach.:2018 IEEE 16th Int. Conf. on Dependable, Autonomic & Secure Comp., 16th Int. Conf. on Pervasive Intelligence &Comp., 4th Int. Conf. on Big Data Intelligence & Comp., and 3rd Cyber Sci. & Tech. Cong.

Koteshwara, S., & Das, A. (2017). Comparative study of authenticated encryption targeting lightweight IoT applications. *IEEE Design & Test, 34*(4), 26.

Krawczyk, H., & Eronen, P. (2010). HMAC-based extract-and-expand key derivation function (HKDF). Internet Engineering Task Force (IETF) Request for Comments (RFC 5869), https://tools.ietf.org/html/rfc5869.

Martínez, V. G., Encinas, L. H., & Dios, A. Q. (2015). Security and practical considerations when implementing the elliptic curve integrated encryption scheme. *Cryptologia, 39*(3), 244–269. https://doi.org/10.1080/01611194.2014.988363.

McGrew, D., & Viega, J. (2005, May). The Galois/Counter Mode of operation (GCM). Submission to NIST.

Montgomery, P. (1985). Modular multiplication without trial division. *Mathematics of Computations, 44*, 519–521.

Moosavi, S. R., et al. (2016). End-to-end security scheme for mobility enabled healthcare Internet of Things. *Journal of Future Generation Computer Systems, 64*, 108.

Okello, W. J., Liu, Q., Siddiqui, F. A., & Zhang, C. (2017). *A survey of the current state of lightweight cryptography for the Internet of things*. IEEE international conference on computer, information and telecommunication systems (CITS).

Patel, M., & Wang, J. (2010). Applications, challenges, and prospective in emerging body area networking technologies. *IEEE Wireless Communications, 17*(1), 80–88.

Raza, S., Helgason, T., Papadimitratos, P., & Voigt, T. (2017). SecureSense: End-to-end secure communication architecture for the cloud-connected internet of things. *Future Generation Computer Systems*. Elsevier, *77*, 40.

Sandoval, M. M., & Uribe, C. F. (2005). *A hardware architecture for elliptic curve cryptography and lossless data compression*. IEEE International conference on electronics, communications and computers, pp. 113–118.

Satoh, A., Sugawara, T., & Aoki, T. (2007). High-speed pipelined hardware architecture for Galois counter mode. *Information Security*, 118–129.

Suárez-Albela, M., et al. (2019). Clock frequency impact on the performance of high-security cryptographic cipher suites for energy-efficient resource-constrained IoT devices. *Sensors, 19*(1), 15. https://doi.org/10.3390/s19010015.

Tadesse Abebe, A., et al. (2019). Efficient FPGA implementation of an integrated bilateral key confirmation scheme for pair-wise key-establishment and authenticated encryption. In F. Zimale, T. Enku Nigussie, & S. Fanta (Eds.), *Advances of science and technology. ICAST 2018. Lecture notes of the Institute for Computer Sciences, Social Informatics and Telecommunications Engineering* (Vol. 274). Cham: Springer.

Ting, P.-Y., Tsai, J.-L., & Wu, T.-S. (2017). Signcryption method suitable for low-power IoT devices in a wireless sensor network. *IEEE Systems Journal, 12*, 2385.

Ullah, S., Li, X. -Y., & Zhang, L. (2017). *A review of signcryption schemes based on hyper elliptic curve.* IEEE, international conference on big data computing and communications.

Wanderley E. et al. (2011). Security FPGA Analysis. In: Badrignans B., Danger J., Fischer V., Gogniat G., Torres L. (eds.) *Security trends for FPGAs* (pp. 7–46). Springer, Dordrecht.

Wu, H., & Preneel, B. (2013). AEGIS: A fast authenticated encryption algorithm. Selected Area in Cryptography SAC.

Xilinx Vivado Design Suite User Guide High-Level Synthesis. UG902 (v2018.3) (December 20, 2018). http://www.xilinx.com/support/documentation/sw_manuals.

Yalla, P., & Kaps, J. P. (2017, Dec). Evaluation of the CAESAR hardware API for lightweight implementations. In *International conference on reconfigurable hardware (ReConFig 2017)* (pp. 1–6).

Data Analytics for Security Management of Complex Heterogeneous Systems: Event Correlation and Security Assessment Tasks

Igor Kotenko, Andrey Fedorchenko, and Elena Doynikova

1 Introduction

Information technology is an essential part of operation of almost any modern organization. But along with the obvious advantages, its application leads to the information security risks and, consequently, to the potential serious material losses. To avoid them the information security management process should be implemented in an organization. According to the standards, the goal of information security management consists in selection of appropriate security control measures that should protect information assets and ensure the credibility among the participants of information exchange.

To achieve this goal, the security management systems on the market implement, in one or another form, the following tasks and components. First, the raw security data, represented in different formats, is entered in a data collection component from internal and external sources. Then the data is processed in a data correlation component. All gathered data are stored in a hybrid data storage. Next, the processed data is entered in a security assessment component and the results of assessment go to a decision support component. A visualization component allows one to represent reports on a current security state and suggestions from the decision support component. Then the decisions from the decision support component, that can be approved by the user via the visualization component if necessary, go to a component responsible for enforcement of these decisions.

I. Kotenko (✉) · A. Fedorchenko · E. Doynikova
St. Petersburg Institute for Informatics and Automation of the Russian Academy of Sciences, St. Petersburg, Russia
e-mail: ivkote@comsec.spb.ru; fedorchenko@comsec.spb.ru; doynikova@comsec.spb.ru

S. K. Shandilya et al. (eds.), *Advances in Cyber Security Analytics and Decision Systems*, EAI/Springer Innovations in Communication and Computing,
https://doi.org/10.1007/978-3-030-19353-9_5

The task of the efficient (i.e., adequate, timely and commercially viable) security management is non-trivial. It is still not solved in spite of numerous research and specific commercial proposals on the market. The challenges in security management are most typical for the organizations that have the following characteristics: large information infrastructure, including a large number of interconnections between its components; distribution; heterogeneity; dynamism, i.e., rapid modifications of infrastructure together with modifications in software and hardware, modifications of the security policy, implementation of the security controls, etc.; openness, i.e., flexible facilities to connect to the system from outside. These challenges consequently lead to a large attack surface (the ways in which the system can be compromised), the intricacy of infrastructure specification, the difficulty in determination of primary and secondary assets and calculation of their criticality, the complexity to detect modifications and security incidents, and to trace the causes of incidents, and the difficulty in calculation of the collateral damage when implementing security controls.

From our point of view, to overcome the aforementioned challenges, it is necessary to develop an efficient approach to integrated analysis of big heterogeneous security data. This approach will form the basis of the data correlation technique. The security data, in this case, refers to information specifying the objects of the analyzed infrastructure, their behavior, and interaction.

Data analysis is one of the most complicated and important tasks in many areas, including security management. Therefore, this chapter considers both the currently existing and proposed data analysis methods. It also describes their use in the selected security management tasks, namely, data correlation and security assessment.

Thus, the purpose of this chapter is to form a methodological basis for data analysis in security management, as well as to demonstrate its practical application, using the event logs from the Windows operating system and from the SCADA power management system.

The chapter is organized as follows. Section 2 explains main aspects which we emphasize in the chapter related to the main theoretical and practical results. Section 3 covers the main related works in the field of security management, correlation of security events, security assessments, and data mining. Section 4 considers the process of security management as a whole, and the suggested approaches to security event correlation and security assessment. The proposed approach to security event correlation, including the developed models, methods, and techniques, is described. Our approach to assessing overall security based on a Bayesian attack graph and CVSS is provided, as well as new methods for determining and assessing asset criticality, and assessing the probability of attack success and predicting the attack development. Section 5 demonstrates the application of the methods and approaches described in Sect. 4 for correlating data of logs of the Windows operating system and of a SCADA power management system. At the end of the chapter, we present conclusions on the application of data analysis methods in security management tasks and their prospects.

2 The Peculiarities of Event Correlation and Security Assessment

The *event correlation* allows one to detect security incidents, as well as the chains of security events that led to these incidents. In the chapter we consider the correlation process as a whole, and such its key stages as normalization, aggregation, and directly correlation (other stages, namely, filtering, anonymization and prioritization are out of scope of this chapter), as well as the main methods of data correlation and security data sources.

Existing security management systems usually implement a rule-oriented approach to event processing. The disadvantages of using this approach consist in its complexity and large time costs required to determine the rules manually. In addition, the effectiveness of correlation, when applying this approach, directly depends on the qualification of administrator, who determines the rules.

The data mining is preferable for the efficient security management because it allows correlating unconditional events with minimal manual settings. The approach proposed in the chapter is based on a syntactic and semantic analysis of security events and information. It is designed to implement the process of adaptive data correlation in large-scale heterogeneous uncertain infrastructures.

The uncertainty of an infrastructure is determined by the absence of any previously known information about its architecture, the type of its elements, their characteristics and relationships.

The large scale of the infrastructure is determined by the conditionally unlimited number of information sources and their types. The key feature of the approach is definition of various relationships between the properties of events within the automated adaptation of the correlation process. The result of analysis of the obtained relationships is a type and structural definition of the analyzed infrastructure, or its approximation, including detection of the most stable connections between its elements. In the chapter, we describe the model of uncertain infrastructure and the techniques of correlation of security events and information.

The *security assessment* task is related to identification, analysis and evaluation of the security risks. The chapter describes our approach to the security assessment based on the Bayesian attack graphs (that represent all possible attack paths in the system and their probabilities) and open security data representation standards, including common platform enumeration standard, common vulnerabilities and exposures standard, and common vulnerability scoring system (CVSS).

We outline advantages and disadvantages of this approach and explain the reasons and advantages of the transition to the approach based on the security data mining. In particular, in the risk analysis scope, the risks should be quantified. This task is divided into the task of calculating impact in case of successful attacks, and the task of calculating the probability of successful attacks.

To calculate possible impact, first of all, it is necessary to determine the assets of information system and calculate their criticality. These tasks are often solved manually by the experts. The manual determination of the infrastructure of information system and its primary and secondary assets is complicated, especially if the infrastructure is constantly changing. Therefore, for its automation, we suggest using the correlation and data mining methods described in the chapter, including the above-mentioned approach to overcoming the uncertainty of infrastructure by the syntactic and semantic analysis of the events occurring in it.

Assessing the criticality of assets is also complicated by the subjectivity and dependence on the qualification of experts, as well as by the dependencies of functioning of some assets on others, which are easy to miss in manual assessment, and which can dynamically change. From our point of view, the aforementioned approach for syntactic and semantic analysis of events will allow solving this problem by automated determination of the interrelations between infrastructure objects (or assets), and by building their hierarchy reflecting the frequency of using various assets. That is, the automation of this process through the use of data analysis methods will improve the accuracy and efficiency of assessments of the assets criticality and, as a consequence, of possible impact.

Calculation of the probability of attack success is also a non-trivial task, depending largely on the quality of input data. In the proposed approach, an attack is a sequence of attack actions consisting in exploitation of the software and hardware vulnerabilities. We calculate the probability based on the CVSS expert assessments of vulnerabilities. CVSS is one of the mostly used and efficient means for vulnerability assessment. We use it to resolve the problem of absence of incident statistics to calculate the probability of attack using the Bayesian approach. Further, this probability can be refined based on the processed incidents. But CVSS assessments and this approach have some disadvantages. In the chapter, we describe these disadvantages and our idea how to overcome them using data mining for vulnerability assessment. We also describe our approach to calculation of the probability of attack success considering incident statistics.

The latter is a part of the dynamic security assessment, that is, monitoring changes in the security state over time taking into account the events occurring in the analyzed system.

We consider two situations. The first situation takes place when a new security incident has been detected. In this case, the event correlation methods are used to identify the entire chain of events that preceded the incident, in order to subsequently train the system to forecast the similar incidents. The second situation takes place when the previously observed incident has been detected. In this case, the methods of attack development forecasting based on the Bayesian attack graph are used. It is important to stop the attack in time, before the essential damage, and to collect the maximum amount of information about an attacker at the same time. The chapter describes the proposed approach, including the criteria for determination of moment to implement the countermeasures to stop the attack. With accumulation of the statistics on security incidents and the resulting chains of security events, the prediction accuracy can be improved.

We demonstrate operation of the proposed correlation techniques using two data sets. One of these includes data logs of the Windows operating system, and the second – data logs of the SCADA power management system. Using both data sets, we show the application of data normalization and correlation techniques. As a result, the main events, properties and objects of the Windows operating system and SCADA system are determined.

3 Related Work

Security management task as a whole incorporates the tasks of information security development, implementation, operation, monitoring, analysis, maintenance and enhancement. Security management systems are based on the application of risk assessment methods for this goal (ISO/IEC 27001:2013 2013). Therefore, this chapter discusses risk assessment methods for security management. Besides, when exploiting security management system, it is necessary to detect and prevent security incidents. Therefore, these tasks are considered in details in the chapter.

The functionality of assessing security risks, as well as processing and analyzing security events from various sources, is implemented by security information and event management (SIEM) systems. Generally, SIEM systems implement the following functions: gathering of data on security events from various sources for forensics and compliance with regulatory acts; normalization to represent the records of events from various sources in an unique format for event analysis; correlation to connect records of events and events from various systems and applications to speed up the detection and response to security threats; aggregation to reduce the amount of data on events by removing identical records; and reporting to present the results to interested parties in real time or in the form of long-term reports.

By performing the listed functions, the SIEM systems can improve the efficiency of security management in an organization by simplifying the process of analyzing events, their timely detection, and prompt processing disparate events. Besides, SIEM systems may simplify the security audit process by addressing the issue of security information consistency and the process of identifying perpetrators by storing event records.

Currently there is a large number of commercial SIEM solutions, including IBM Security (QRadar SIEM 2019); HP (ArcSight SIEM 2019); Splunk (Splunk Enterprise Security 2019); LogRhythm (LogRhythm SIEM 2019); IntelSecurity (IntelSecurity SIEM 2019).

During the operation of SIEM systems, the raw security data is collected in the data collection component. Then the data is processed in the data correlation component. All gathered data are stored in the hybrid data storage. Next, the correlated data enter the security assessment component and the results of assessment go to the decision support component. The visualization component represents reports on the current security state and suggestions from the decision support component. Then the decisions from the decision support component, that can be approved by the user via the visualization component if necessary, go to the component responsible for application of these decisions (MASSIF 2019). This description shows that the component of correlation and data processing, as well as the component of risk assessment, are among the main ones in the security management system. Therefore, as the main tasks, we highlight the correlation of security data and the security assessment based on risk assessment.

Correlation is a statistical dependency between random variables, that cannot be expressed with functional dependency (Hazewinkel 2001). This dependency can be determined using parametric and nonparametric correlation indices including linear, various rank and other. In particular, there are Pearson coefficient (SPSS Tutorials: Pearson Correlation 2019), Spearman's rank correlation coefficient (Wayne 1990a), Kendall rank correlation coefficient (Wayne 1990b), Goodman and Kruskal's gamma (Goodman and Kruskal 1954), etc.

Correlation of data is used in many information security tasks, including evaluation of algorithms, detecting patterns of distributed DoS attacks, identifying subsets of data attributes for intrusion detection (Beliakov et al. 2012; Jiang and Cybenko 2004; Wei et al. 2013). For example, it is applied in intrusion detection systems (IDS) for detection of attacks based on the relationships between network events (Kruegel et al. 2005). In this chapter, we consider the process of security management using SIEM systems. Therefore, most attention will be paid to the event correlation in SIEM systems.

Many modern SIEM systems use a rule-based correlation (Hanemann and Marcu 2008; Limmer and Dressler 2019; Müller 2009). It is based on a fixed correlation of events under certain conditions that may contain logical operations on data, their properties and calculated indicators. The main advantage of this method is ease of implementation. Its main drawbacks are as follows: the complexity and long time for computing the rules by the security administrator; its efficiency directly depends on the skills of administrator; and the system that uses such method is not able to detect the incidents if such rules were not initially specified for them. There are also methods based on templates (scenarios) (Hanemann and Marcu 2008), graphs (Ghorbani et al. 2010; Xu and Ning 2008), finite state machines (Hasan 1991; Xu and Ning 2008), similarity (Gurer et al. 1996; Zurutuza and Uribeetxeberria 2004) and others, but they can also be expressed in the form of rules.

Currently, the methods that should eliminate these disadvantages are being developed. They are based on the intelligent data analysis. It refers to self-learning approaches such as Bayesian networks (Bursztein and Mitchell 2010; Dantu et al. 2009; Fedorchenko et al. 2017), immune networks (Tiffany 2002; Xu and Ning 2008), artificial neural networks (Elshoush and Osman 2011; Kou et al. 2012; Tiffany 2002; Xu and Ning 2008) and others. For example, in (Jiang and Cybenko 2004) a probabilistic event correlation model for network attack detection based on spatial-temporal analysis is proposed. Here event spaces are linked in a chain of sequences, and a concrete state from the set of states corresponds to each space at the current moment. The resulting chain sequences are used to calculate the probability of a specific attack scenario. Davis et al. (2017) use the application behavior model to identify illegitimate and abnormal activity. The initial data for constructing the model are events that reflect the system calls of all possible applications. Based on the initial information, a normal behavior profile is formed by processing a multi-graph where each vertex is an event.

The advantage of these approaches is in the possibility of an independent (unconditional) event correlation with minimization of manual settings. Their disadvantages are the complexity of learning model building, additional requirements to the adequacy and quality of the models, and to the completeness of original training data.

Security assessment is rather broad and complex task, the purpose of which is to provide sufficient information to determine the need to implement security control measures in the analyzed system and to select the set of countermeasures. This process includes identification, analysis and evaluation of security threats that can lead to an unacceptable damage (impact) for the considered organization.

The unacceptable damage is specified differently for different organizations. In particular, for mission-critical facilities, the probability of successful implementation of any security threats should be close to zero. An unacceptable damage level for the commercial organizations depends on the financial losses that organization can afford to incur while eliminating the damage. For example, in (Lockstep Consulting 2004) a scale for governmental organizations is proposed. It includes the following levels of damage that are outlined considering the efforts that should be spent on recovery in case of successful implementation of threats (from lowest to highest): €0 – negligible (almost complete absence of damage in case of successful threat implementation, additional recovery costs required); €1000 – small (small damage to the value of the asset, almost no additional recovery costs required); €10,000 – significant (perceptible but small damage, requires some recovery costs); €100,000 – damaging (damage to the reputation and/or resources of the organization, requires significant recovery costs); €1,000,000 – severe (system failure and/or loss of customers or business partners, costs equal to the cost of full restoration of resources); €10,000,000 – grave (complete compromise and destruction of the organization, the annual budget of the organization is required for restoration).

The final assessment may be presented in the form of an integral system security metric (for example, based on the integral level of security risk or an attack surface), or in the form of a list of threats with risk level values.

When using a risk level metric for the security assessment, this process is directly connected with risk assessment and incorporates risk identification, risk analysis and risk evaluation (ISO/IEC 27005:2008 2008).

The first risk identification stage consists in the identification of main entities that participate in the security management process, including assets of the analyzed system, their owners, possible threats and their sources and ways of their implementation including vulnerabilities of the analyzed system, implemented in the analyzed system security measures and tools. One of the risk identification methods is manual listing of threats, assets, etc. based on the questionnaires or criteria, for example, in the form of tables. This method is used, for example, in scope of the techniques "Facilitated Risk Analysis and Assessment Process" (FRAAP) (Peltier 2010) and "Operationally Critical Threat, Asset, and Vulnerability Evaluation" (OCTAVE) (Caralli et al. 2007). The main advantage of this method is ease of implementation. Its main drawbacks consist in that it is time consuming, human-intensive and subjective. In the conditions of modern, heterogeneous, distributed and large-scale systems, these disadvantages are essential. This led to the development of a variety of methods based on intelligent data analysis.

In particular, for assets identification, the tools of manual inventory of assets are used, such as in ISMS (Heron 2019), which implies adding objects to the database when they appear. But for this approach, the above disadvantages are especially relevant. Besides, there is an approach that consists in using XML tags to identify software (ISO/IEC 19770-1:2017(en) 2017). But in practice, it is impossible to limit all software installed in open systems, and use only software supplied with tags. And it is also impossible to track all the software of newly connected mobile devices or Internet of Things (IoT) devices, software from additional electronic media. In addition, this applies only to software products.

Besides there are techniques for automated discovery of information system objects. Network scanners are used to discover such objects as hosts, software and services, and their interconnections (NMap 2019; Tenable 2018; Wireshark 2019). Also, there are techniques for automated detection of service dependencies (Agarwal et al. 2004; Brown et al. 2001; Ensel 2001; Hanemann 2007). But these tools and methods may determine only limited number of objects. For a detailed risk assessment and an adequate security assessment, it is important to determine other critical system objects (for example, files), to assess the importance of various objects, the connections between the various objects of system over time, to determine the rights of various users in the system, on which the possibility of transition from one attack action to another (i.e., from one host to another) will depend, among other things, as well as to highlight the malicious object (process, session, file), that is, to determine initially "dynamic" objects, changing over time.

There are some approaches in this area based on data mining (Tuchs and Jobmann 2001). In (Tuchs and Jobmann 2001) the frequency characteristics of event types are used to discover the objects. In (Motahari-Nezhad et al. 2011) the event sequences are analyzed to discover the objects. In this chapter, we describe our technique for discovering dynamic objects of uncertain infrastructure based on the dynamic data (event) analysis to have an actual dynamic structure of the information system that is close to (Motahari-Nezhad et al. 2011; Tuchs and Jobmann 2001). Unlike (Tuchs and Jobmann 2001), we use not only frequency characteristics of event types to discover the objects, but also the utilization rate and variability of properties and their values. It allows more accurate determination of objects. Unlike (Motahari-Nezhad et al. 2011), we started from the hierarchy of objects and their types, and moved to analysis of the event sequences of separate objects on the next step. From our point of view, it will allow more accurate comparison of event sequences, division of the events by relation to the objects, and determination of connections between the objects. Our task is to determine infrastructure objects by monitoring events in the system. The accuracy of such determination depends, among other things, on how complete the logging of events is, and in addition, a lot of preliminary work is required on parsing the event logs of different systems.

To identify the threats, in addition to manual listing of threats and vulnerabilities, which can be supplemented by penetration testing, various approaches are also proposed, including those based on game theory, where the game of an attacker and a defender (strategy) is defined as a sequence of states and actions (that lead to transition to new states) (Bursztein and Mitchell 2010), on system map (a dependency graph of priority resources) (Balepin et al. 2003), on attack trees and attack graphs (sequences of all possible attack actions and transitions between them) (Dantu et al. 2009; Lippmann et al. 2007; Noel et al. 2003; Poolsappasit et al. 2012).

One of the problems when identifying potential threats is the problem of zero-day vulnerabilities. Penetration testing and reverse engineering of software and hardware are used to solve it. But their quality directly depends on the qualifications of the expert and requires huge time costs. In the last decade, a method for analyzing the behavior of software in a virtual environment (the sandbox) has been actively developed to detect an abnormal or malicious behavior. But such approach just reduces the number of instances for the further manual analysis.

Besides, there are solutions related to monitoring the state of system during its operation in order to detect security incidents based on analysis and correlation of events (Müller 2009; Sadoddin and Ghorbani 2006). But existing solutions are limited with application of the signature and rule-oriented approaches, and usually do not allow detecting and preventing cyberattack timely. A promising direction in development of event correlation methods is the use of data mining methods (Müller 2009; Sadoddin and Ghorbani 2006; Tiffany 2002). Advantage of these approaches consists in the opportunity of self-dependent (unconditional) event correlation with minimization of manual setting. But to construct training models, a preliminary analysis of the data itself and training of classifiers is required, and the initial training data must be sufficiently complete.

Known publications on security assessment, taking into account zero-day vulnerabilities, are few in number and mainly use subjective metrics (Joshi and Singh 2018; Wang et al. 2014). Besides, there are metrics that reflect the possibility of zero-day attacks (Ahmed et al. 2008).

Risk analysis can be based on the qualitative expert assessments as in the mentioned above techniques FRAAP (Peltier 2010) and OCTAVE (Caralli et al. 2007). Besides, there are various metrics and techniques of quantitative security assessment. In case of the security assessment based on the risk assessment, two main metrics are used – the probability of successful attack implementation and the impact level in case of successful attack implementation.

The probability of successful attack implementation can be calculated directly based on attack trees and attack graphs. In particular, in (Dantu et al. 2009; Poolsappasit et al. 2012) the probability (or potentiality) of each attack action is calculated on the basis of the Bayesian attack graph. Besides, the frequency of attacks of a certain type per year can be used to calculate this probability (Hoo 2000).

There are different approaches to calculate the attack impact (or damage). In a number of studies, it is related to a loss of functionality after the attack. For instance, in (Wu et al. 2007) the impact depends on the number of affected by the attack system operations and security goals. In (Jahnke et al. 2007) the attack impact is calculated as the percentage of resources that are available after the attack. In other research works, the attack impact is directly connected with the criticality of assets under the attack. In (Hoo 2000) the losses from attacks are calculated considering the observed losses from attacks of a certain type in the past. In (Toth and Kruegel 2002) the impact that is specified with the penalty cost metric depends on the importance of the attacked service and reduction of its productivity. In (Balepin et al. 2003) the attack impact is calculated on the basis of a system map (a dependency graph of priority resources) and a cost model (to assign costs for the resources manually). The attack impact is calculated as the sum of costs of resources (dependency graph nodes) affected by the attack. Besides, there are approaches with a partially automated impact determination, when the criticality of the core assets is set manually and the criticality of the rest assets is calculated automatically using service dependency graphs (Kheir et al. 2010; Wu et al. 2007).

In addition to assessing security based on risk analysis, there are other approaches. For example, some of them are based on the resources that can be compromised (Lippmann et al. 2007) or the attack surface (when probabilities are not considered) (Manadhata et al. 2007).

In the chapter, we describe our multilevel approach to security assessment that is based on the use of common vulnerability scoring system (CVSS) and other open standards, the available data, and how to refine this assessment when new data appears. The approach has the following features: unified representation of the input data on the base of open standards; joint consideration of the characteristics of different objects of assessment (network hardware and software, vulnerabilities, attacks, attackers, events and countermeasures); application of service dependency graphs and Bayesian attack graphs to calculate metrics; hierarchical division of metrics on groups according to the data used for calculations (Doynikova and Kotenko 2017). We distinguish security assessment techniques for the static and dynamic modes of system operation.

4 The Proposed Approaches to Security Event Correlation and Security Assessment

The security management incorporates a set of tasks of different type. Their automated implementation is assigned to SIEM systems. These tasks consist of collecting security data and its processing, including correlation, security assessment, and decision support. Event correlation and security assessment are key tasks of security management, because ultimately the quality of security decisions depends on them.

Figure 1 depicts connections between these tasks and their subtasks.

We propose an approach to *event correlation* that is based on the structural analysis of the security events for determination of the direct (absolutely equivalent) links between event properties and construction of a graph of event types on their basis (Fedorchenko et al. 2017). The advantage of this approach is the automated adaptation of the correlation process to uncertain target infrastructures.

Correlation process is continuous and should be executed in the real time. The place and role of correlation process in SIEM systems are determined with such tasks as: (1) revealing the interrelations between heterogeneous security informa-

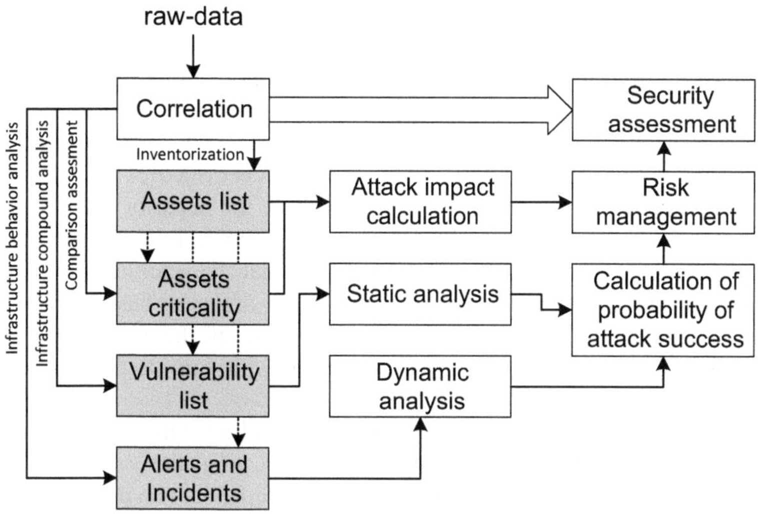

Fig. 1 Interconnections between the correlation and security assessment tasks

tion; (2) determining the information objects of the targeted infrastructure and their characteristics; (3) grouping the low-level events in the high-level meta events; and (4) detecting the incidents and security alarms on the basis of behavior analysis of infrastructure objects of different levels. Thus, correlation process performs data processing starting from their coming from heterogeneous sources to generation of the report on the current security state.

The common approach to *security assessment* includes the following stages: (1) input data collection (risk identification stage of the risk assessment), assets inventory in Fig. 1 relates to this stage; (2) calculation of security metrics (risk analysis stage), assets criticality assessment, vulnerability assessment and security incidents processing and analysis in Fig. 1 relate to this stage; (3) definition of security level (risk evaluation stage).

On the first stage, data from different sources (experts, network scanning tools, open vulnerability databases, SIEM systems, open databases of weaknesses, and attack patterns) is gathered and processed. The data is gathered in the unified format of the SCAP protocol (Radack and Kuhn 2011), and we generate our own models and structures for further analysis (Kotenko and Doynikova 2018):

- The analyzed system model based on assets inventory (for the assets inventory we propose to use methods of correlation)
- The service dependency graph based on data specifying the system services and their dependencies (Kheir et al. 2010) (it is used to calculate the asset criticality)
- The attacks model represented as a Bayesian attack graph based on the network model, vulnerabilities and CVSS indexes (Poolsappasit et al. 2012)
- The security event model based on security incidents (from the SIEM system) that are generated using our approach to events correlation

On the second stage, the generated models are used for risk evaluation based on the taxonomy of security metrics (Kotenko and Doynikova 2014). This taxonomy classifies security metrics according to the input data and the stages of security analysis (static and dynamic). It incorporates the following levels: topological (metrics related to the analyzed system configuration and its objects); attacks level (metrics related to the possible attacks); attacker level (metrics that characterize attacker); security events level (metrics related to the security events and incidents); countermeasure (decision support) level (metrics related to the countermeasure); integral level (includes integral metrics that characterize system security as a whole). Three sub-categories for each category are used: base characteristics, cost characteristics (calculated with consideration of the monetary value of resources), and zero-day characteristics (defined with consideration of zero-day vulnerabilities).

We used existing metrics that were redefined according to the requirements, standards and models used for calculations. This classification allows us to differentiate security assessment techniques for different levels depending on the available input data. For example, assets criticality is calculated on the topological level using a service dependency graph. The prior probability of attacks is calculated on the attacks level (in static mode). To calculate prior attack probability considering the local probabilities of separate attack actions and the conditional probabilities of attack actions as part of sequences, we assess vulnerabilities of the system. We outline two possible alternatives – use of CVSS and use of data (exploits) analysis. On the events level, the posteriori probabilities of attacks are calculated considering correlated incidents (in dynamic mode). Each level of metrics allows one to assess the security situation. This assessment can be improved with new data (from other levels) by applying the techniques of the appropriate level of metrics to recalculate the metrics of the previous level and to calculate new metrics.

On the third stage, the security level is defined (as the set of security metrics). Common security assessment is represented using metrics of the integral level. To calculate metrics of the integral level, it is necessary to have at least metrics of the topological level as input data.

Let us describe below the methods and techniques used on each level.

4.1 Security Data Correlation

Figure 2 depicts a generalized scheme specifying the relations of diverse types of actions and different infrastructure assets taking into account the defense lines. The scheme illustrates hypothetical relations between the main participants of the cyber-physical interaction.

We divided the sources of actions into anthropogenic (employees, customers and potential security violators in each of these roles), technogenic (technological and IT processes), and the external environment (natural physical environment). These sources of actions can also be active and passive, external and internal assets.

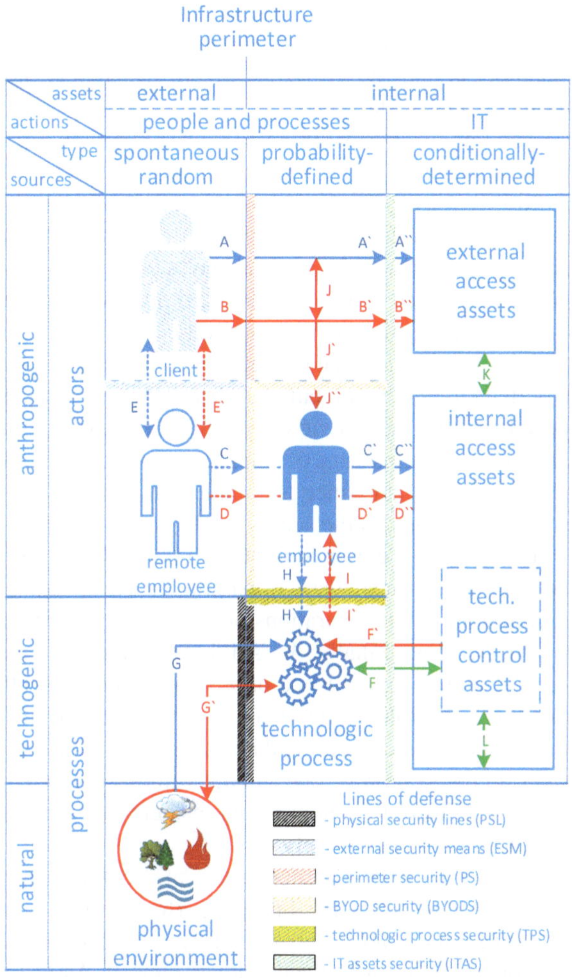

Fig. 2 The general scheme of relations between impact sources and infrastructure assets across the lines of defense

We also identified three main types of actions that affect infrastructure: spontaneously random, probability-defined, and conditionally-deterministic. The actions of the latter type are represented in green (*F*, *K*, *L*), while the blue and red colors of the relationships reflect the nature of the influence: legitimate and malicious actions, respectively, in terms of security.

Consider the boundaries of infrastructure protection in more detail. Despite the fact that IT assets are divided into separate groups, each of the defense lines can contain both physical (fence, door, lock, etc.) protection measures, and means (software and hardware, computer network) ensuring cybersecurity.

That is why in most cases the passage of actions through the next line of defense is complemented by the impact on the line itself. The exception is the impact of the physical environment on the line of physical protection. This fact is due to the impossibility of guaranteed protection against natural disasters (earthquake, hurricane, tsunami, etc.). Despite the fact that customers and employees are also assets (in a certain sense), the main assets that require security are the technological process and the IT infrastructure to support business processes (interaction with customers, staffing).

The line of physical protection (*PSL*) and the line of ensuring the safety of the technological process (*TPS*) should be noted, since they are bidirectional. In other words, we must control the implementation of the technological process not only by automated security tools and people, but also guarantee the limitation of possible consequences when the process leaves control. In this regard, the actions of I' and G' are performed by the technological process on the employees who provide it and the external environment, respectively.

It is worth noting that the security tools used in the above protection lines are not strictly belong to them, but there is a rather strict separation of their functional purpose from the point of view of security. For example, the perimeter protection (*PS*) also controls the passage of employees to the target infrastructure. However, it is not taken into account, since the implementation of the check mode on the territory strongly depends on the access model used.

Let us take a closer look at the relationships shown on the scheme. Impact A defines the legitimate ways of passing the perimeter security line (*PS*). Similarly, A' specifies legitimate operations of infrastructure client relative to the IT assets security (ITAS) line. Finally, A'' is a direct impact on external access IT assets. In the case of malicious influence on *PS* and *ITAS* or *BYODS* (Bring Your Own Device Security), the role of the client becomes the role of the attacker, and it should be calculated when correlating the data and taken into account when assessing security risks. In the same way, one can determine legitimate (C, C', C'') and malicious (D, D', D'') actions of employees and insiders (internal violators), respectively, including those working remotely.

We pay a special attention to the interaction of employees and customers (E and E'). In the case when an employee performs working actions remotely, the only line of protection is *ESM* (external security means), the responsibility for the implementation of which lies with the employee himself. As E, for example, it can be a private phone call, personal email, or informal meetings. Malicious actions of E' can be directed both against the employee (social engineering) and against his/her digital devices (without contact with the employee). In this case, even if such actions can be partially automatically recorded, they can hardly be guaranteed to be extracted and used for analysis and the subsequent correlation of information. Thus, poorly controlled actions on the scheme are indicated by a dotted line (including H, H', I, I').

Due to application of more serious security measures to external sources of impacts on the infrastructure (clients and external attackers) than to internal (employees), passing the boundary of the perimeter protection (*PS*) in a legitimate or improper manner can be used to further penetrate into main assets. These actions are indicated in the diagram as J, J' and J'' and allow one to expand the number of possible combinations of sources of influence on the infrastructure used in the attack. For example, the sequence of actions $AJB'B''KF$ will lead to the implementation of legitimate actions to control the process by third-party users (if security policy allows this), while $AJB'B''KF'$ – can lead to disastrous consequences G'.

We also explain the impact G of the external environment on the technological process. Examples of such effects can be the natural physical phenomena, such as wind, turning of water in a river, and nuclear reaction for such technological processes as generation of energy with the help of wind turbines, hydroelectric power stations and a nuclear power plant, respectively. In this case, the most critical task from the point of view of safety is the correct separation of the environmental impacts G and G' on the technological process.

The presented scheme of the effects of heterogeneous sources on infrastructure assets is a generalized, abstract and universal representation of real actions occurring in the target infrastructure. In real infrastructures, separate lines of defense and the corresponding activity of the impact initiators may be unacceptable a priori.

However, using this scheme, we can describe the organization of the security of any cyber-physical systems, including the IoT and automated process control systems. Such a presentation is focused on the preliminary correlation of conditionally determined types of actions for the subsequent assessment of the presence of probability-defined and spontaneously random impacts, as well as the nature of their impact on the target infrastructure. Each action taken in relation to a certain line of events, or between technogenic sources of influences is recorded in the event logs.

Source data for the structural analysis is a set of security events E in a log L.

$$E^L = \{e_1, e_2, \ldots, e_n\}.$$

An event is understood as a fact or result of some action on any of its stages: an attempt (failure event), start of action (start event), intermediate result of an action (lengthy events), the result of action (completed correctly, completed with error).

Each event e comprises a set of tuples of property p and its value v:

$$e = \{(p, v)\}, \quad p \in P, \quad v \in V, \quad P = \{p_1, p_2, \ldots, p_d\}, \quad V = \cup_{i=1}^{d} V^{p_i},$$

where P and V^p – a set of properties and a set of their values, accordingly, and d – a total number of properties. At each point in time these sets are finite and not empty, though they can change their cardinality in time:

$$P \neq \varnothing, \; V^{p_i} \neq \varnothing, \; \forall i \in [1; d].$$

Each property p of event e is defined by a value v from the set of possible values V^p.

Thus, the main definitions that describe initial data for the correlation process are provided. For the further description of the approach to events correlation on the basis of structural analysis we should introduce a model of uncertain infrastructure and its connection with the initial data.

At each point in time, an uncertain infrastructure I consists from the set of information objects O:

$$O^I = \{o_1, o_2, \ldots, o_s\}.$$

These objects exist in time (i.e., they have some lifetime), and their characteristics and states are specified using some characteristics x. Information objects are necessarily connected via possession relations. It means that each object is necessarily a part of a higher-level object and (or) contains lower-level objects. Also, the connection between objects is determined considering their direct interaction with each other.

We suppose that the set of characteristics X^o, that describes an information object o, uniquely identifies the type of information object ot.

Each characteristic x from the set X belongs to only one type of the information object ot.

Proceeding from the above, the formal statement of the particular task of this study is the development of an approach f, that should transform initial data of the correlation process to the sets of objects and their types of the uncertain infrastructure:

$$\{E, P, V\} \xrightarrow{f} \{O, OT\}$$

Let us assume that the state of the target infrastructure I is specified with the events E in the log L. The connection between the security events E and the infrastructure objects O is based on the following statement: each property p of the event e is a characteristic x of the object o.

In other words, each event describes at least one information object of the infrastructure. Each event e has a specific set of properties $P^e \subset P$, that describe actions.

We assume that different actions produced by the set of objects O are specified by different types ET of the events E. It means that each type et is unique (it doesn't repeat among other types) in terms of combination of the properties P^{et}:

$$ET = \{et_1, et_2, \ldots, et_q\},\ et = P^{et},\ P^{et} \neq \varnothing,\ P^{et} \subset P,$$

$$P^{et_i} \neq P^{et_j}\ \forall i, j \in [1; q], i \neq j,$$

where q – a number of event types.

Structural analysis of security events is based on the research of their properties and connections between the properties. We suggest dividing the set of types of relation between properties on (1) same type and (2) inter types. The properties of the same type can be equivalent in two features: (1) absolute similarity and (2) semantic (type) similarity. The inter types properties can be equivalent in different features: uniqueness, equiprobability, usability, etc. Each type of relations between security events properties forms the set of appropriate types of links that is denoted by character ~ with specification of the equivalence feature of the relation type: $p_1 \overset{abs}{\sim} p_2$.

It means the connection of the property p_1 with the property p_2 on the basis of absolute equivalence, or that property p_1 is absolutely equivalent to the property p_2. Thus, specific types of relations between properties determinate the appropriate characteristics of properties relative to each other. $f^{feat}()$ denotes the functions for determination of connections between any two properties p_1 and p_2, where *feat* – an equivalence feature of the relation type implementing the connection between properties.

Below in this study we consider each researched type of relations between properties separately to perform structural analysis.

Same type relations based on absolute similarity between the event properties e_1 and e_2 mean similarity of their types if each property of event e_1 has this connection with property of event e_2, and vice versa:

$$et^{e_1} = et^{e_2} \Leftrightarrow \forall p^{e_1}, p^{e_2} \in P : \exists\ p^{e_1} \overset{abs}{\sim} p^{e_2}.$$

If this condition is satisfied only for one event and is not satisfied for another, then there may be the hierarchy of event types. But this study doesn't consider this aspect of analysis.

Thus, the function for determination of the same type connections by the absolute similarity feature f^{abs} is mapping of the events set E and the set of their properties P on the set of event types ET. It can be specified as follows:

$$f^{abs}\left(e^1, e^2\right) = \begin{cases} P^{e_1} \Delta P^{e_2} = \varnothing : \ et^{e_1} = et^{e_2}; \\ \text{otherwise} : \ et^{e_1} \neq et^{e_2}. \end{cases}, \ f^{abs} : \{E, P\} \rightarrow ET,$$

where Δ – an operator of symmetric difference of sets. This mapping is surjective because for each element of the set ET its complete inverse image is not an empty set:

$$\forall\, et \in ET : \left\{ f^{abs-1}\left(et\right) \right\} \neq \varnothing$$

The same type relations between properties are introduced for the typification of events. It will allow analyzing behavior of the separate information objects on the basis of their sequences of event types in the future.

It should be noted that the result of comparing the types of two events by absolute similarity is a binary, i.e., event types and their properties are equivalent or not. All other types of relations define the characteristic of the connection between properties in quantitative equivalent.

We assume that each event property relates to some type, and the meaning of each property is unique. But the meaning of some properties can coincide. *Same type relations based on type equivalence* between the event properties specify the set of the type properties PT:

$$PT = \left\{ pt_1, pt_2, \ldots, pt_m \right\},$$

where m – a number of event property types. The function for determination of such connections f^{type} is based on the following hypothesis: the intersection of sets of values V^{p_1} and V^{p_2} for two properties p_1 and p_2 of the same type, that are equivalent in terms of types similarity, is not an empty set:

$$V^{p_1} \cap V^{p_2} \neq \varnothing \ \forall p_1 \overset{type}{\sim} p_2,\ \{p1, p2\} \subset P,\ \left\{V^{p_1}, V^{p_2}\right\} \subset V$$

Ideally, an intersection of sets of values V^{p_1} and V^{p_2} of the same type properties p_1 and p_2, that are equivalent in terms of types similarity, is equal to each set V^{p_1} and V^{p_2}. The function of determination of such type of relations between properties is specified as follows:

$$f^{type}(p1, p2) = \frac{\left|V^{p_1} \cap V^{p_2}\right|}{\left|V^{p_1} \cup V^{p_2}\right|}.$$

Thus, this function is the surjective mapping of the set of the properties P and the sets of their values V in the set PT, because the set of all inverse images for each type of the properties pt is not empty:

$$f^{type} : \{P, V\} \rightarrow PT,\ \forall pt \in PT : \left\{f^{type-1}(pt)\right\} \neq \varnothing.$$

If the property p doesn't have the equivalent by type similarity properties in P, then the set of its values forms separate semantic property type.

Intertype relations between security event properties can be specified by different features, such as: mutual use of properties in events and use by types of events, the mutual uniqueness of property values, the possibility of correlation of property values, and many others. But the proposed approach for the structural analysis of events considers only *intertype relations* between the properties, that determine the properties *equivalence on the basis of their mutual use*.

We assume that different security events E describe different actions of objects-sources with objects-goals from the set O. The objects are specified using the object characteristics X. The function of determination of these intertype relations f^{using} is based on the following hypothesis: exclusive-sharing of the properties p_1 and p_2 defining characteristics x_1 and x_2 of one or more objects from the set O indicates the belonging of the described objects to one or more object types OT. We also assume that absolute rate of use of properties indicates the level of the described objects in the hierarchy of types of infrastructure objects.

The function of determination of inter-type relation based on the mutual use f^{using} between event properties p_1 and p_2 is specified as follows:

$$f^{using}(p_1, p_2) = \frac{\left|E^{p_1} \cap E^{p_2}\right|}{\left|E^{p_1} \cup E^{p_2}\right|}, \{p_1, p_2\} \subset P, \left\{E^{p_1}, E^{p_2}\right\} \subset E.$$

Considering the above, the function f^{using} is a mapping of the event set E and the set of their properties P to the set of object types OT:

$$f^{using}:\{E,P\}\rightarrow OT.$$

It should be noticed that the strength of links between the events has range of values [0,1] if their equivalence is given by the real index. Therefore, a threshold *lim* should be specified to establish the existence of connection between properties (transfer to a Boolean result). The overcoming of the threshold allows considering the link valuable. This parameter was analyzed in details for the intertype relations between properties considering an equivalence of their use. It is described in the results section below.

Thus, we described the approach to security events correlation based on their structural analysis. The approach is developed to connect heterogeneous information and detect information objects of the uncertain target infrastructure.

We made a few assumptions to perform structural analysis of the event types. These assumptions concern content of the events log, the information objects of the target infrastructure and their relations. The developed approach has some requirements to the source data, namely:

1. Any type of events *et* can't contain two absolutely equivalent properties:

   $$\forall\{p_1,p_2\}\in P^{ot}:p_1\not\sim p_2,$$

   because in this case, it is impossible to determine belonging of the characteristics x to the described objects O uniquely.

2. The event properties characterize the specific information type uniquely, values of each property have uniform presentation format, and the semantics of two properties of different types cannot coincide. Otherwise, in case of use of non-normalized data, application of the approach will lead to significant number of false negatives while determining the same type relations between properties based on the type equivalence.
3. Completeness of the initial data should be sufficient to decide whether there is a link between the properties of events. It means that each event type *et* should correspond to at least one event E. Otherwise, the event type *et* doesn't exist. Also each property p should exist in at least one event type ET, and the set of values V^p of each property P is not empty. Otherwise, the property p doesn't exist.

4.2 Security Assessment

The following general features of our approach to security assessment should be noted. The approach involves the use of all available security data, both at a fixed point in time and during the further exploitation of the system. In accordance with this, we distinguish the risk assessment techniques for the static and dynamic mode of system operation.

The approach focuses on a quantitative security assessment, and it is based on the use of a number of security metrics of different levels (topological, attacks, attacker and events (Kotenko and Doynikova 2014)). The risk assessment techniques for different modes of system operation use different sets of security metrics. Thus the technique for the static mode of system operation uses only metrics of the topological, attacks and attacker levels. The technique for the dynamic mode of system operation uses also metrics of the events level.

4.2.1 Input Data Collection

As it was mentioned above input data collection relates to the risk identification stage. The security assessment results depend directly on the quality and completeness of the analyzed data. On this stage we identify main entities that participate in the security assessment process and generate our own models and structures for further analysis. We use data from various sources for their specification, including experts, various scanners, open databases, system security logs, intrusion detection, and intrusion prevention systems etc. These data can be divided on static (software and hardware, vulnerabilities gathered by network scanners) and dynamic (obtained, for example, from intrusion detection and prevention systems).

The list below represents main entities that participate in the security assessment process, the data used for their specification, and sources of these data:

- *Assets entities* – We identify them by determining the configuration of the analyzed system (including objects of analyzed system and links between them) and further specification of the assets. To identify them, we propose two approaches. The first approach involves the use of expert knowledge and network scanners to generate the model of the analyzed system and the service dependency graph. The second approach is based on the correlation technique to identify objects of the analyzed system and their interrelations. The sources for their identification are experts, network scanning tools, "National Vulnerabilities Database" (NVD) (NVD 2019a) as source of software and hardware unique identifiers according to the "Common Platform Enumeration" (CPE) standard (NVD 2019b), "Common Vulnerability Reporting Framework" (CVRF) (ICASI 2019), and "Common Configuration Enumeration" (CCE) database (Common Configuration Enumeration (CCE) 2019) for specification of the infrastructure.
- *Sources of threats* – Currently we identify only one source of threats – attacker – and we describe it with the set of attacker characteristics. The sources of data for attacker identification are expert knowledge (in the static mode of system operation) and security events and incidents that allow one to extend the knowledge on the attacker in the dynamic mode of system operation.

- *Possible threats and ways of their implementation* – We describe possible threats as an attack graph where each node corresponds to the vulnerability/weakness exploitation or implementation of attack pattern. The sources for their identification (except analyzed system configuration) are databases of vulnerabilities (namely, NVD (NVD 2019a)), weaknesses (namely, "Common Weakness Enumeration" (CWE) database (Common Weakness Enumeration 2019)) and attack patterns (namely, "Common Attack Pattern Enumeration and Classification" (CAPEC) database (Common Attack Pattern Enumeration and Classification 2019)) and also the results of the technique that we propose to discover zero-day vulnerabilities. The technique is based on the analysis of the exploits code from "Exploit DataBase" (EDB) (Exploit Database 2019), determining the characteristics of exploits connected with known vulnerabilities, connection of the determined characteristics with information on the events in real time to fix exploitation of zero-days.
- *Security measures and tools implemented in the analyzed system* – We describe them with the set of their characteristics. The sources for their identification are expert knowledge, network scanning tools and available checklists of security controls (for example, from NVD).

On the base of the gathered data, we generate the models that are used in the techniques described below.

Consider the process of automated assets inventory in details. Our technique of the assets inventory, based on the correlation, incorporates several consequent stages: (1) gathering and preliminary processing of the input data on events from the information system logs; (2) determining objects (assets), object types and hierarchical connections between them on the basis of the statistical analysis of the gathered data.

We use correlation method to determine object types and objects themselves. Each event is specified with the set of characteristics (event properties), including characteristics that specify the object to which the event refers. Each type of information objects has its own set of characteristics, which uniquely identifies it. Examples of object types: "process" (service), "file", "session" (network, user), "sensor", "host" (network equipment, personal computer), etc.

As soon as characteristics can have the same data type but different semantic nature (for example, "SourceAddress" and "DestAddress"), we propose to use the metric of pair variability of their values to determine the same characteristics first. The formula for calculating this metric is different for static characteristics (i.e., characteristics whose values on average change much less frequently than values of dynamic characteristics, for example, the process name) and dynamic characteristics, such as process identifier. In the case of static characteristics, the metric of attribution of characteristics to the same type is determined in the same way as for typing properties of events (getting of the property types set PT is described earlier).

In the case of dynamic characteristics, types are determined by analyzing the lifetime and variability of the values of the properties on this segment.

To outline the groups of characteristics, probably characterizing the types of objects, we determine the metric of their pair use. The pair use of characteristics means the ratio of the cases of their (pair) common use to the total number of their observations in events. Calculation of the metric of characteristics pair use is performed by equation of pair use of event characteristics, which is described earlier (getting the set of object types OT).

To determine the configuration of the analyzed system, in addition to the objects, it is necessary to determine the connections between them. Each object is a necessarily part of a higher-level object and (or) contains lower-level objects, that is, it is a part of an object hierarchy.

We calculate the total use of object characteristics to determine the level of objects in the hierarchy (the higher the use, the higher the object in the hierarchy). It is calculated as a ratio of the number of uses of the object characteristic in the events to the total number of all events. The relationships between object types are specified by the types of events containing them. There are two main types of relationships: object state change and interaction between objects. Existence of the interaction between the objects defines the existence of connection between the object types.

4.2.2 Calculation of Security Metrics

This stage corresponds to the risk analysis stage. For this goal, we use our taxonomy of security metrics (Kotenko and Doynikova 2014) and developed techniques for static and dynamic modes of system operation. Let us describe the technique for static mode of system operation first. We calculate risks for the identified attacks. We define attack as a sequence of attack actions. Each action is exploitation of the vulnerability of software or hardware. We define risk according to the existing standards as product of the attack probability and its impact on the system operation (Doynikova and Kotenko 2017): $Risk = AttackImpact \bullet AttackPotentiality$, where *AttackImpact* – attack impact; *AttackPotentiality* – attack probability. In its turn, we calculate attack impact on the basis of affected assets criticality and attack damage.

The damage from an attack action a_i that exploits vulnerability v_i against asset R_k is specified on the basis of the CVSS indexes as the vector $[ConfImpact_{k,i}(c) \quad IntegImpact_{k,i}(c) \quad AvailImpact_{k,i}(c)]$, where $ConfImpact_{k,i}(c)$ – damage level for the confidentiality c of R_k in case of successful implementation of a_i, $IntegImpact_{k,i}(i)$ – damage level for the integrity of R_k, $AvailImpact_{k,i}(a)$ – damage level for the availability of R_k. We specify possible values for $ConfImpact_{k,i}(c)$, $IntegImpact_{k,i}(i)$ and $AvailImpact_{k,i}(a)$ as {0.0; 0.275; 0.660} according to the CVSS indexes *confidentiality impact*, *integrity impact* and *availability impact* (Doynikova and Kotenko 2017).

In the following section calculation of attack criticality and attack probability is described in details.

Assets Criticality Assessment

The asset criticality metric is defined for the asset's R_k confidentiality $cCrit_k$, integrity $iCrit_k$ and availability $aCrit_k$.

We propose the techniques that implement two different approaches to calculate asset criticality. The first technique requires rather significant manual work, and it is based on the application of the service dependency graph.

It includes the following stages (Doynikova and Kotenko 2017):

1. Determine the criticality of the business assets considering business losses in case of violation of asset confidentiality, integrity and availability (CIA). This stage involves close cooperation with business owners.
2. Determine the criticality of secondary assets (i.e., information technology (IT) assets that support business assets): (1) determine the IT assets that directly support business assets; (2) calculate criticality of these primary IT assets; (3) calculate criticality of the IT assets that support primary IT assets; (4) calculate the criticality of these secondary IT assets.

Asset criticality can take values from 0 to 100: [0:0.01) – Insignificant; [0.01:0.1) – Minor; [0.1:1) – Significant; [1:10) – Damaging; [10:100) – Serious; 100 – Grave (based on the (Lockstep Consulting 2004)).

The second technique is based on the application of the correlation methods and includes the following steps: (1) determination of the object type place in the object types hierarchy, that is the part of assets identification process described in the previous Sect. 5.2.1; (2) calculation of the relative criticality of the asset considering its place in the hierarchy and in relation to the objects of the same type.

In step (1), the object place in the hierarchy depends on the value of metric of its total use. In step (2), we suggest to use the metric of total use to determine critical infrastructure objects (assets). This metric allows determining the most used objects. Thus, object criticality level relative to other types depends on its place in the hierarchy (the higher the object in the hierarchy, the higher its use, and, accordingly, the higher its relative criticality). The various intervals of values of this metric correspond to the scale of values of criticality given above.

It should be noted that the proposed approach allows automated calculating the relative criticality of infrastructure objects. But further improvements and refinements are required including by determining the dependencies between objects and the associated effect on the criticality of objects. Besides, in this case, rarely used critical objects are not taken into account. It will be considered in future research.

Finally, returning to the attack impact metric (*AttackImpact*), we multiply damage level of an attack action a_i and the calculated criticality of the affected asset R_k to calculate *AttackImpact*:

$$AttackImpact = cCrit_k \times ConfImpact_{k,i}(c) + iCrit_k \times IntegImpact_{k,i}(c) + aCrit_k \times AvailImpact_{k,i}(c)$$

Probability Assessment

The prior probability of attacks is calculated on the attacks level (in static mode) using attack graph. It is defined as the set of nodes (attack actions) and links between them. Links between nodes define the opportunity to transfer from one attack action to another. We link two nodes depending on pre-conditions and post-conditions of an attack action and network connections between hosts.

The likelihood of transfer from one node to another is defined on the base of discrete local probability distributions for nodes. There are two main relation types between linked nodes: AND – to compromise the node, it is necessary to compromise all parent nodes (in our graph it is sequential nodes); OR – to compromise the node, it is necessary to compromise at least one parent node (Kotenko and Doynikova 2018).

It is the so-called Bayesian attack graph (Dantu et al. 2009; Poolsappasit et al. 2012). It allows considering influence of events on the system security state, forecasting attack development and outlining previous attack steps.

Our technique for calculation of an attack probability (*AttackPotentiality*) for attack graph nodes uses and evolves previous works on the Bayesian attack graphs (Frigault et al. 2008; Poolsappasit et al. 2012). Main distinctions in our approach are in the method of an attack graph generation and in the algorithm of calculation of local probabilities of compromise for the graph nodes.

To calculate prior attack probability (*AttackPotentiality*) considering local probabilities of separate attack actions and conditional probabilities of attack actions as part of sequences, we use the technique that incorporates the following stages (Doynikova and Kotenko 2017): (1) calculation of the local probabilities of compromise for the graph nodes; (2) calculation of the discrete conditional probability distributions; and (3) calculation of the unconditional probabilities.

In stage (1), we calculate local probabilities of compromise for the graph nodes using vulnerability metrics, namely, CVSS index *Exploitability* (*Exploitability* = 20 × *AccessVector* × *AccessComplexity* × *Authentication*, where *AccessVector* determines required access to exploit the vulnerability, *AccessComplexity* determines complexity of the vulnerability exploitation and *Authentication* defines if exploitation of the vulnerability requires additional authentication) as in (Poolsappasit et al. 2012). We redefine equation for the *Exploitability* to calculate a local probability for node S_i that corresponds to the attack action a_i considering that the attack graph is constructed so that transition from one state to another is possible only if there is access to the appropriate graph node:

$$p(a_i) = 2 \times AccessVector \times AccessComplexity \times Authentication,$$

If $S_i \in S_r$, where S_r – root nodes of the graph.

If $S_i \notin S_r$ local probability for node S_i is calculated as follows:

$$p(a_i) = 2 \times AccessComplexity \times Authentication.$$

Discrete conditional probability distributions $Pc(S_i|Pa(S_i))$ for the graph nodes (i.e., probabilities of compromise of S_i considering different combinations of states of its parents $Pa(S_i)$) are defined on the basis of the reverse depth-first traversal for the graph considering "AND" (to compromise child node it is necessary to compromise all parent nodes) and "OR" (to compromise child node it is necessary to compromise at least one parent node) relations between linked nodes. Nodes of the attack path are connected with "AND" relation by construction, while "OR" relation forms different paths.

We calculate discrete conditional probability distributions using equations from (Frigault et al. 2008):

$$Pc\left(S_i|Pa\left(S_i\right)\right)=\begin{cases}0, & \exists S_i \in Pa\left(S_i\right) \mid S_i = 0\\ p\left(S_i\right) & \text{otherwise}\end{cases}\text{, in case of "AND" relations;}$$

$$Pc\left(S_i|Pa\left(S_i\right)\right)=\begin{cases}0, & \forall S_i \in Pa\left(S_i\right) \mid S_i = 0\\ p\left(S_i\right) & \text{otherwise}\end{cases}\text{, in case of "OR" relations.}$$

Final prior probabilities (unconditional probabilities) for the graph nodes are calculated on the basis of local probabilities and discrete conditional probability distributions as follows:

$$\Pr\left(S_1,\ldots,S_n\right)=\prod_{i=1}^{n} Pc\left(S_i|Pa\left(S_i\right)\right).$$

The technique for the dynamic mode of system operation is directly connected with events analysis and processing based on the correlation of the events into incidents.

Security Incidents Processing

The posteriori probabilities of attacks are calculated considering correlated incidents (in dynamic mode) on the events level of the proposed metrics taxonomy.

We introduce incident model *ev* to use information on the security incidents for security assessment.

It includes the following fields: attack target (it can be account, process, data, component, host, or network); attack result (it can be privileges escalation, information disclosure, information violation, denial of service, resources consumption).

Besides, additional information can be considered, if exists: attacker characteristics; attack goal, etc. (Kotenko and Doynikova 2018).

The events model allows connecting information on the security incidents with models of previous levels of the taxonomy (including the attack graph and the attacker model). It allows mapping the attacker position on the attack graph and calculating security metrics that reflect the system security state in the dynamic mode of system operation (including the attacker position in the system, attacker skills, past and future attack steps).

Mapping is based on the object of the analyzed system where incident was fixed and vulnerabilities of this object that lead to the privileges and\or the impact specified in the incident.

Mapping of the incident on the attack graph node(s) leads to recalculation of probability that the node(s) is compromised (*AttackPotentiality*). The techniques on the basis of Bayesian attack graphs are applied for recalculation (calculation of the posteriori probabilities) (Dantu et al. 2009; Poolsappasit et al. 2012). The probability that node S is compromised, i.e., probability of success for the attack action a, is recalculated as follows (Kotenko and Doynikova 2018):

$$p(a|ev) = \frac{p(ev|a) \times p(ev)}{p(a)} = \frac{p(ev|a) \times (p(ev|a) \times p(a) + p(ev|-a) \times p(-a)}{p(a)},$$

where

$p(a)$ – the probability of node compromise before security incident;

$p(ev|a)$ – the information reliability index for an incident ev that determines probability that incident ev is true, this index depends on the source of information on the incident, in our case, it is reliability of our correlation technique;

$p(ev|-a)$ – the probability that ev is false (false positive).

The probabilities for the child nodes of attacks that go through the compromised node are redefined taking into account the new probability of the compromise for the node for which the security event happened.

Besides, new knowledge on the attacker can be obtained from the events and incidents analysis.

The main stages of techniques for the calculation (recalculation) of the attacker skills are as follows (Kotenko and Doynikova 2018):

1. Mapping of security incidents on the graph.
2. Recalculation of the attack action probability based on the Bayes' theorem.
3. Recalculation of the probabilities of ancestors of the compromised node to define the most probable attack path to this node. The probability that the parent node b was attacked if the child node a is compromised:

$$p(b|a) = \frac{p(a|b) \times p(b)}{p(a)},$$

where $p(a|b)$ – the prior probability of compromise of child node if the parent node was compromised (the conditional probability that was defined on the base of the attack graph); $p(b)$ – prior probability of compromise of the parent node (the unconditional probability that was defined on the base of the attack graph);

$p(a)$ – the probability of compromise of the child node that was defined on the previous step.

4. Selection of the vulnerabilities with the maximum CVSS access complexity index (Mell et al. 2007) for the processed path. We define the attacker skill level according this access complexity (the value on the scale High/Medium/Low corresponding to quantitative values 0.7/0.5/0.3).

Recalculated attacker skill level further is used in multiple tasks, including recalculation of the attack probabilities, redefinition of the attack goals, etc.

4.2.3 Definition of Security Level

This stage corresponds to the risk evaluation stage. We use the risk level calculated on the previous stage to determine common security assessment.

Security assessment is based on the risk level for the whole information system. In its turn, the risk level for the whole information system is calculated on the basis of the calculated risk levels for the attack graph nodes: at first we calculate the risk level for an attack as combination of the minimum probability for the attack actions constituting this attack and the maximum impact, then we calculate the risk level for system objects as the maximum risk for attacks that go through appropriate object; and finally we calculate the risk level for the information system as the maximum risk for the objects (Doynikova and Kotenko 2017).

The risk can take values from 0 to 100, where the risk from 0 to 0.1 is low that corresponds to the high security level (risk can be accepted), the risk from 0.1 to 1 – medium that corresponds to the medium security level (countermeasures should be implemented), the risk from 1 to 10 – high that corresponds to the low security level (countermeasures should be implemented as soon as possible), and the risk from 10 to 100 – critical that corresponds to very low security level (countermeasures should be implemented immediately) (Kotenko and Doynikova 2018).

5 Experiments and Discussion

We implemented and tested the approach proposed in the previous section for correlating security information for individual security assessment tasks on the following datasets: (1) Windows 8 security event log (we specifically created this dataset for experiments); (2) the US Energy Management System (EMS) event log at an investor owned utility (description and dataset No.5 are available at (EMS 2017a), the log contains anonymized events recorded for the SCADA system from mid-2017).

The characteristics of the analyzed logs are presented in Table 1.

A special feature of the EMS dataset is the unformalized presentation format (syslog analog) of the computer-generated event records, in contrast to the

Table 1 Characteristics of the experimental datasets

Dataset characteristics	Windows 8	EMS
Source data size	4 Gb	1.4 Gb
Log format	Evtx	Syslog
Duration of recording	1 month	1 month
Number of events	6,700,000	5,758,500
Number of real event types	44	52
Number of possible event types	250	69
Number of formalized event properties	110	9

formalized Windows log. In this case, by a formalized representation, we understand the named event properties and the corresponding values.

We used the EMS set only for the experiment on the automated identification of event types, since in order to carry out typing of event properties and assets, the non-formalized data requires additional analysis in order to identify syntactic structures and official words. In turn, the Windows security event log dataset was also used to identify event property types and infrastructure object (asset) types.

5.1 *Identification of Event Types*

To maintain the purity of the experiment on analyzing the event logs of uncertain infrastructures, we initially excluded some properties from the EMC dataset. These properties are defined in the description and title. Thus, in the EMC dataset, we left unnormalized (semantically undefined) event records of the *event_message* field. In the entries of this field, we also ignore the first three words, since they duplicate the event properties defined in (EMS 2017b): *Substation*, *DeviceType*, *Device*. The authors of the EMC dataset initially anonymized the names of operators, devices, and objects. To test the functionality of the proposed event typing method, we used the SCADA_Category field, which has 69 possible values stated in the description of the dataset. However, in this field we observe 52 unique values corresponding to the types of event records.

Based on the proposed approach, the typing of events according to their structure, the number of event properties (and their names) in the records must be identical for events of the same type. In the case of non-formalized data, the number of tokens (lexemes) is a sign of the initial division of events into types.

Thus, we broke the dataset into 21 types, in which each event record had from 4 to 24 tokens. Further, we presented all the properties in each type of event as a tuple from the type number (by the number of tokens in the record) and the sequence number of the tokens (counting from the beginning of the record).

This step is required to separate event types with the same number of tokens. The subsequent analysis consists in identifying the properties of events containing lexemes of various data types (not the same as the semantic type of properties). Thus, we presented each type of events in the form of sequences of lexical data types, namely: date, integer value (int), fractional value (float), percents (integer or fractional value with symbol "%" at the end) and nominal (qualitative) values – tokens whose type does not belong to any of the other types. From the 21 event types, 14 have mixed properties.

Below we present examples of similar sequences consisting of the first letters of the specified data types:

- iqqqqqq
- qqqqqqf
- qqqqqqq
- iqqqqqqqqfqf
- qqqqqqqqqqqq

Thus, we correctly identified 52 observed types of events. It is worth noting that in the case of typing events by analyzing unformalized data, difficulties may arise with an unambiguous interpretation of various lexical types even within the same journal. That is why the proposed approach is robust, but not automatic. Various variants of collisions are also possible, when an unambiguous definition of the type of events due to their additional structural division by type. However, this research direction we plan to consider in further work.

In the process of log analysis, we detected 37 unique sets of properties. Each of them specifies the separate event type. Further analysis showed that seven pairs of the declared types have equal sets of used properties.

The correspondence of these types is presented in Table 2.

On the one hand, the obtained results indicate the difference in the types of events from the variants of the set of their properties. On the other hand, the number of event types describing related actions, i.e., the actions of the same types of objects, have an equal sets of properties. As a result of detection of the same type relationships between the event properties on the basis of absolute equivalence, these types were combined into one common. Thus, the empirical accuracy of detection of event types using the proposed approach to correlation of Windows 8 dataset security events is 84%. In future research, we plan to analyze the opportunity of determination of events types based on the values of their properties.

5.2 *Identification of Property and Assets Types*

As a result of analysis of the initial data, 41 properties were classified on 13 types. Let us specify the null type that represents null properties.

The property is null if it takes a single value that does not characterize the property meaning. This type is needed in case of the insufficient data normalization when values of some properties cannot be filled correctly for certain events. Thus, all properties of the null type for all processed events get null value. In our case, this value is the "–" character, and a quantitative index of types equivalence for each pair from the set of null properties is equal to 1. In the described initial data there

Table 2 The Correspondence of event types with the equal sets of properties

Event type 1	Event type 2
4670: Permissions on an object were changed	4907: Auditing settings on an object were changed
4778: A session was reconnected to a Windows Station	4779: A session was disconnected from a Windows Station
4800: The workstation was locked	4801: The workstation was unlocked
4904: An attempt was made to register a security event source	4905: An attempt was made to unregister a security event source
4946: A change has been made to Windows Firewall exception list. A rule was added	4948: A change has been made to Windows Firewall exception list. A rule was deleted
5154: The Windows filtering platform has permitted an application or service to listen on a port for incoming connections	5158: The Windows filtering platform has permitted a bind to a local port
5156: The Windows filtering platform has allowed a connection	5157: The Windows filtering platform has blocked a connection

are six null properties: *TransmittedServices*, *CommandLine*, *FileName*, *TaskContentNew*, *LinkName*, and *Conditions*.

For each of other types of properties (except the null type), two properties equivalent in type are assigned. The example of outlined properties and their interpreted types (if possible) are provided in Table 3. One of the types could not be interpreted, since the relationship between its properties is unclear.

Figure 3 shows the *DomainName* type of properties generated using links by types equivalence with specification of the threshold f^{type}. Graph nodes denote the properties of event types, and arcs – the weighted relationships between properties indicating the quantitative ratio of the relationship. Links, the indicator of typical equivalence of which is close to 0, are not displayed.

The relations between properties are clearly divided into strong and weak. Obviously, in case of specification of the significance threshold $f^{type} > 0.3$ for these relations, this type would be divided on three additional property types: (1) *ClientName*, *Workstation* and *WorkstationName*; (2) *SubjectDomainName* and *TargetDomainName*; (3) *AccountDomain*.

The determination of object (assets) types by properties, equivalent in mutual use, is based on the hypothesis that the same usability of event properties indicates the description of the characteristics of one or more types of objects.

Figure 4 shows a histogram of using properties relative to the total number of events. Only those properties are shown whose usability in log events exceeds 3%, that is, if the property occurs in less than 3% of events, then this chart does not consider it. It is clearly seen that individual groups of properties have an equal, or very close in value, utilization rate. Thus, the proposed hypothesis is preliminarily confirmed.

As a result of the experiment on determination of the heterogeneous relationships between the properties on the basis of their sharing indicator, 18 property groups were identified, and their total number is 60.

Table 3 The example of properties connected by type equivalence and their types

Interpreted property type	Properties
RemoteID	RemoteMachineID, RemoteUserID
DomainName	AccountDomain, SubjectDomainName, TargetDomainName, ClientName, WorkstationName, Workstation
HandleId	HandleId, SourceHandleId, TargetHandleId
ProcessName	ProcessName, NewProcessName
LogonGuid	LogonGuid, TargetLogonGuid
ProcessId	ProcessId, NewProcessId, SourceProcessId, TargetProcessId
UserSid	UserSid, TargetUserSid, SubjectUserSid
UserName	AccountName, TargetUserName, SubjectUserName

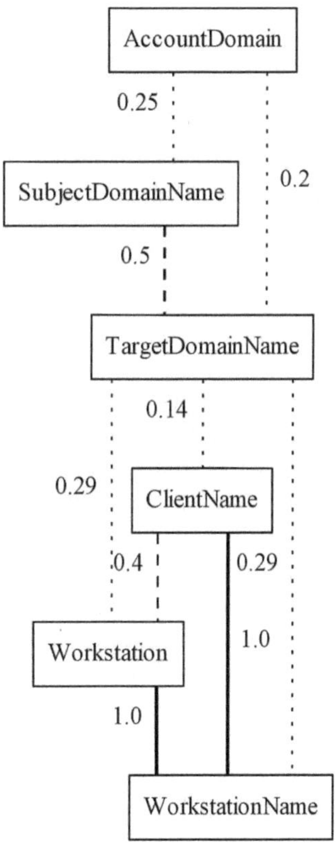

Fig. 3 Properties of DomainName type

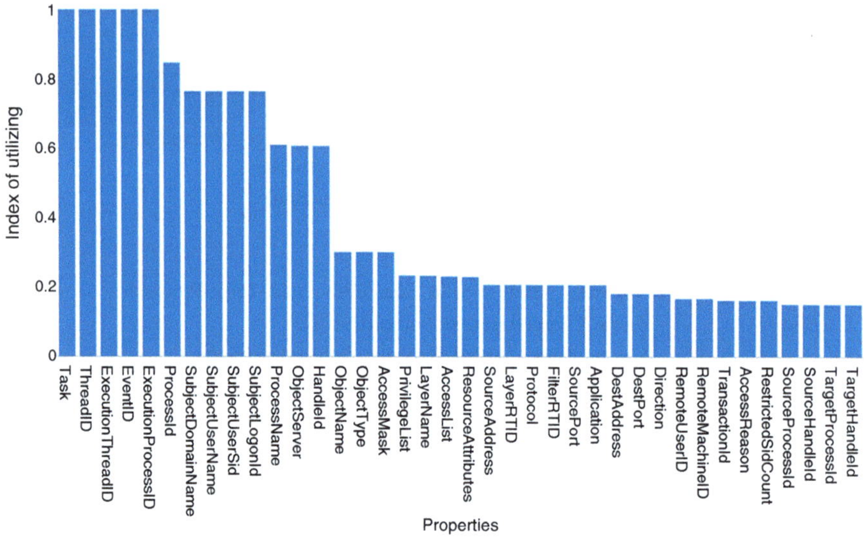

Fig. 4 Histogram of the event properties utilizing in the OS Windows security log

The most significant and interpreted types of objects, together with the properties that define them, are presented in Fig. 5. In this figure, in each group of events, the average value of the general use of the group properties is also noted, and the groups themselves are ordered by this indicator. However, with the threshold of links significance $lim = 1$, the indicated average level of use for the group will correspond to the usability of each property.

The objects of null type deserve a particular attention. They are found in all types of events. Strictly speaking, these properties specify a null event type, since even an empty event as a data structure has a specific header with service information, for example, the event type *EventID* or the *Task*. It is worth noting that the fact of having a null object type (event) in the analyzed infrastructure can be established by checking the utilization rate, since for such properties it should be equal to 1.

Having considered the generated types of objects, we concluded that most of them (with the exception of the null level) are sufficiently semantically expressed. Obviously, the presented property groups form the types of high-level objects.

Thus, groups of properties 1 and 2 form the types of objects “Subject” and “Object”, respectively. At the same time, group 2 clearly should be divided into two types of objects: *Source* and *Target*. Probably, deeper typing of objects, for example, based on their behavior, will allow avoiding such inaccuracies. This direction will be considered in further research.

The obtained object types describe 54% of all available properties of events in the security log with the threshold of links significance $f^{using} = 1$. However, with this threshold value, a number of sufficiently used properties were skipped.

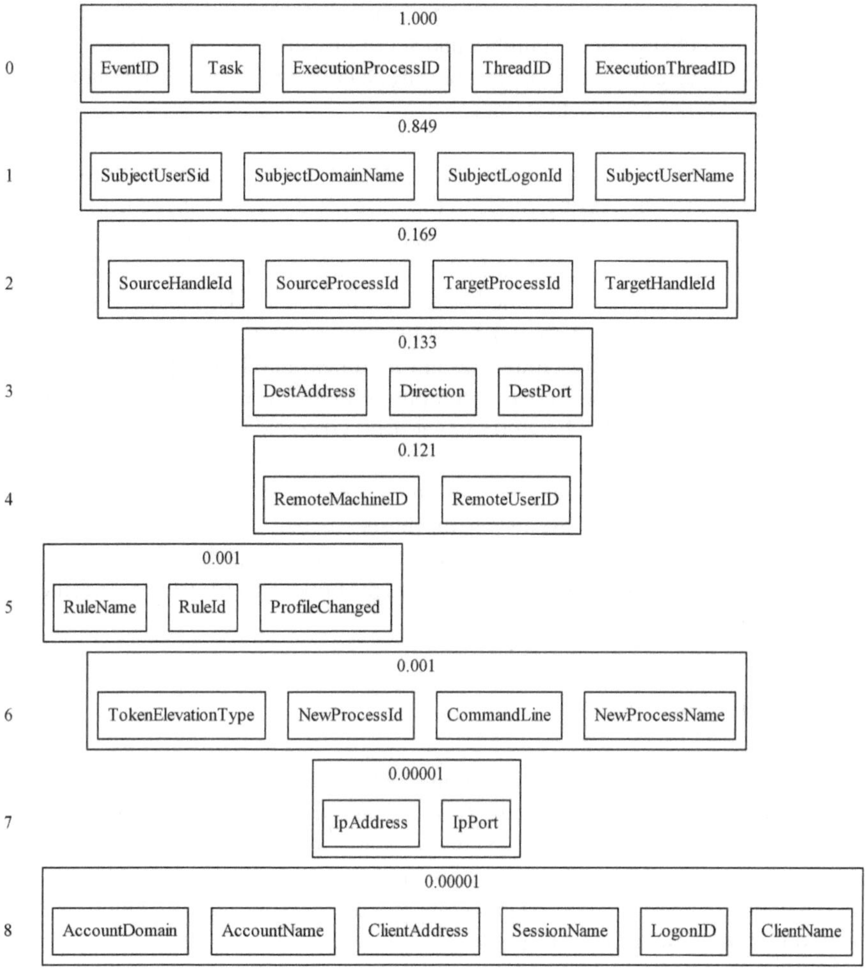

Fig. 5 The example of property groups of object (assets) types in OS Windows

This case is associated with some errors that arise, for example, when the initial data is normalized and when the number of events with a certain normalized property changes insignificantly. Such a change does not guarantee an identical mutual use of properties; therefore, such cases should be handled by means of the threshold of links significance.

In the process of analysis of the threshold of links significance, we found out that even a slight assumption of shared use of event properties leads to an increase in the number of types of objects. When using the threshold of links significance $f^{using} > 0.99$ to determine the relationship between properties by type equivalence, another type of objects was identified, which is defined by three properties: *ProcessName*, *HandleId*, and *ObjectServer*.

The following statements are also correct: changing the threshold of links significance leads to a change in the number and set of properties of object types, as well as to the appearance of collisions. In this case, a collision means assigning a property to several types of objects at once.

The total time for performing the tasks of the described correlation approach is no more than 2 minutes. The longest operations are: (1) loading data into the main memory for further processing, since in this case the download speed is limited by the speed of reading data from the hard disk; (2) definition of equivalence relations on mutual use. The average execution time of both operations is about 30 seconds (the second operation was performed in parallel in six threads). Source code of the prototype is written using Python 3.5 and libraries numpy, scipy, and pandas. The results are visualized using GraphViz and modules matplotlib, pyplot, and seaborn. We used for the experiments a computing platform with one 6-core processor Intel(R) Xeon(R), CPU E5-2603 v3, @ 1.60 GHz and 64 GB RAM.

Given that the initial data are records of events of a sufficiently long period of time, the time of the analysis can be considered acceptable. In general, the proposed approach to correlation of events on the basis of their structural and statistical analysis showed fairly good results, and most of the cases in which it proved unstable are most likely caused by some deviations in the initial data from the requirements.

6 Conclusion

The problems of security event processing and security assessing of complex heterogeneous systems is an increasingly important area of research. Existing data correlation approaches, usually based on rule-based methods, are often not efficient enough for real-time security assessments. In this chapter, we described the models and methods of the proposed adaptive approach for the event correlation and the assessment of the security of IT assets.

The basis of the proposed approach to processing heterogeneous data of uncertain infrastructures is a structural and statistical analysis, which allows one to present security information in an abstract way. As a result of automated typing of events and their properties, as well as the allocation of information asset types, it is possible to significantly reduce the configuration time for the security assessment component. The presented models and techniques were experimentally tested on two different datasets. The results obtained in general confirmed the validity of the hypotheses put forward and the possibility of applying the described approach in practice.

We aimed to describe the heterogeneous data correlation and security assessments. The future research will be devoted to formalization of computer-generated texts (events and security information) for expanding the area of source data suitable for automated correlation, as well as further developing an adaptive semantic description of uncertain infrastructures for the problem of security assessment.

Acknowledgments The reported study was partially funded by RFBR according to research projects No. 16-29-09482 (security metrics and techniques for their calculation), 18-29-22034 (analysis of the techniques based on neural networks), 19-07-00953 (analysis of security assessment techniques), 19-07-01246 (intelligent security data analysis techniques and corresponding experiments), by the budget (the project No. 0073-2019-0002), and the Council for Grants of the President of Russia according to scholarship SP-751.2018.5 (analysis of various sources and subjects of cyberattacks).

References

Agarwal, M. K., Gupta, M., Kar, G., Neogi, A., & Sailer, A. (2004). Mining activity data for dynamic dependency discovery in e-business systems. *IEEE Transactions on Network and Service Management, 1*, 49–58. https://doi.org/10.1109/TNSM.2004.4798290.

Ahmed, M. S., Al-Shaer, E., & Khan, L. (2008). *A novel quantitative approach for measuring network security*. In: INFOCOM proceedings, Phoenix, AZ, USA, April 2008. IEEE, pp. 1957–1965.

ArcSight Enterprise Security Manager (ESM). Security Information and Event Management (SIEM). In: Microfocus Off. website. https://www.microfocus.com/en-us/products/siem-security-information-event-management/overview. Accessed 27 Mar 2019.

Balepin, I., Maltsev, S., Rowe, J., & Levitt, K. (2003). Using specification-based intrusion detection for automated response. In *Proceedings of the sixth international symposium on recent advances in intrusion detection (RAID), Pittsburgh, PA, USA, September 2003. Lecture notes in computer science (LNCS)* (Vol. 2820, pp. 136–154). Berlin, Heidelberg: Springer-Verlag.

Beliakov, G., Yearwood, J., & Kelarev, A. (2012). Application of rank correlation, clustering and classification in information security. *Journal of Networks, 7*, 935–945.

Brown, A., Kar, G., & Keller, A. (2001). *An active approach to characterizing dynamic dependencies for problem determination in a distributed environment*. In: 2001 7th IEEE/IFIP international symposium on integrated network management proceedings: Integrated management strategies for the new millennium, Seattle, WA, USA, May 2001.

Bursztein, E., & Mitchell, J. C. (2010). Using strategy objectives for network security analysis. *Lecture Notes in Computer Science, 6151*, 337–349.

Caralli, R. A., Stevens, J. F., Young, L. R., & Wilson, W. R. (2007). *Introducing OCTAVE allegro: Improving the information security risk assessment process* (Technical report). Software Engineering Institute.

Common Attack Pattern Enumeration and Classification. https://capec.mitre.org/. Accessed 27 Mar 2019.

Common Configuration Enumeration (CCE). In: Off. website. https://nvd.nist.gov/cce/index.cfm. Accessed 30 Mar 2019.

Common Weakness Enumeration. https://cwe.mitre.org/index.html. Accessed 27 Mar 2019.

Dantu, R., Kolan, P., & Cangussu, J. (2009). Network risk management using attacker profiling. *Security and Communication Networks, 2*, 83–96. https://doi.org/10.1002/sec.58.

Davis, M., Korkmaz, E., Dolgikh, A., & Skormin, V. (2017). Resident security system for government/industry owned computers. *Lecture Notes in Computer Science, 10446*, 185–194.

Doynikova, E., & Kotenko, I. (2017). *CVSS-based probabilistic risk assessment for cyber situational awareness and countermeasure selection*. In: Proceedings of the 25th Euromicro international conference on parallel, distributed and network-based processing, St. Petersburg, Russia, March 2017.

Elshoush, H. T., & Osman, I. M. (2011). Alert correlation in collaborative intelligent intrusion detection systems – A survey. *Applied Soft Computing, 11*, 4349–4365.

EMS. (2017a). Energy management system (EMS) events log. In: *Industrial Control System Cyber Attack Datasets*, dataset No.5. https://sites.google.com/a/uah.edu/tommy-morris-uah/ics-data-sets. Accessed 31 Mar 2019.

EMS. (2017b). Description of the energy management system (EMS) events log. In: *Industrial Control System Cyber Attack Datasets*, dataset No.5. https://drive.google.com/file/d/0B8m%2D%2DHm7dOEqRFNOd1hvelZ3ZFE/view. Accessed 31 Mar 2019.

Ensel, C. (2001). *A scalable approach to automated service dependency modeling in heterogeneous environments*. In: Proceedings of the 5th IEEE international enterprise distributed object computing conference, Seattle, WA, USA, September 2001.

Exploit Database. https://www.exploit-db.com/. Accessed 27 Mar 2019.

Fedorchenko, A., Kotenko, I., & El Baz, D. (2017). *Correlation of security events based on the analysis of structures of event types*. In: Proceedings of the 9th IEEE international conference on intelligent data acquisition and advanced computing systems: Technology and applications (IDAACS'2017). IEEE, Bucharest, Romania, pp. 270–276.

Frigault, M., Wang, L., Singhal, A., & Jajodia, S. (2008). *Measuring network security using dynamic bayesian network*. In: Proceedings of the 4th ACM Work Qual Prot. https://doi.org/10.1145/1456362.1456368.

Ghorbani, A. A., Lu, W., & Tavallaee, M. (2010). *Network intrusion detection and prevention concepts and techniques*. New York: Springer.

Goodman, L. A., & Kruskal, W. H. (1954). Measures of association for cross classifications. *Journal of the American Statistical Association, 49*, 732–764. https://doi.org/10.2307/2281536.

Gurer, D. W., Khan, I., Ogier, R., & Keffer, R. (1996). An artificial intelligence approach to network fault management. *Computers and Biomedical Research, 47*, 18–27. https://doi.org/10.3138/carto.47.1.18.

Hanemann, A. (2007). *Automated IT service fault diagnosis based on event correlation techniques*. Dissertation, LMU of Munich.

Hanemann, A., & Marcu, P. (2008). *Algorithm design and application of service-oriented event correlation*. In: 3rd IEEE/IFIP international workshop on business-driven IT management, Salvador, Brazil, April 2008.

Hasan, M. (1991). *A conceptual framework for network management event correlation and filtering systems*. In: Proceedings of the 6th IFIP/IEEE international symposium on integrated in network management, pp. 233–246.

Hazewinkel, M. (2001). Correlation (in statistics). In *Encyclopedia of mathematics* (p. 5402). Cham: Springer Science+Business Media B.V.

Heron, J. *How to develop an Asset Inventory for ISO 27001 – A pragmatic approach*. In: isms.online. https://www.isms.online/iso-27001/how-to-develop-an-asset-inventory-for-iso-27001. Accessed 15 June 2019.

Hoo, K. J. S. (2000). *How much is enough? A risk-management approach to computer security*. Dissertation, Stanford University.

ICASI. *Common vulnerability reporting framework (CVRF)*. In: Off. Website. http://www.icasi.org/cvrf. Accessed 30 Mar 2019.

IntelSecurity SIEM. In: Websecure Off. website. https://www.websecure.com.au/intel-security/security-information-event-management. Accessed 27 Mar 2019.

ISO/IEC 19770-1:2017(en). (2017). *Information technology – IT asset management – Part 1: IT asset management systems – Requirements*.

ISO/IEC 27001:2013. (2013). *Information technology – Security techniques – Information security management systems – Requirements*.

ISO/IEC 27005:2008. (2008). *Information technology – Security techniques – Information security risk management*.

Jahnke, M., Thul, C., & Martini, P. (2007). *Graph based metrics for intrusion response measures in computer networks*. In: Proceedings of the conference on local computer networks, Dublin, Ireland, October 2007.

Jiang, G., & Cybenko, G. (2004). *Temporal and spatial distributed event correlation for network security*. In: Proceedings of the American Control Conference. IEEE Xplore, pp. 996–1001.

Joshi, C., & Singh, U. K. (2018). An enhanced framework for identification and risks assessment of zero-day vulnerabilities. *International Journal of Applied Engineering Research, 13*, 10861–10870.

Kheir, N., Cuppens-Boulahia, N., Cuppens, F., & Al, E. (2010). *A service dependency model for cost-sensitive intrusion response*. In: 15th European conference on research in computer security, pp. 626–642.

Kotenko, I., & Doynikova, E. (2014). Security assessment of computer networks based on attack graphs and security events. *Lecture Notes in Computer Science, 8407*, 462–471.

Kotenko, I., & Doynikova, E. (2018). Selection of countermeasures against network attacks based on dynamical calculation of security metrics. *Journal of Defense Modeling & Simulation, 15*, 181. https://doi.org/10.1177/1548512917690278.

Kou, G., Lu, Y., Peng, Y., & Shi, Y. (2012). Evaluation of classification algorithms using MCDM and rank correlation. *International Journal of Information Technology & Decision Making, 11*, 197–225.

Kruegel, C., Valeur, F., & Vigna, G. (2005). *Intrusion detection and correlation. Challanges and solutions* (Vol. 14). New York: Springer US.

Limmer, T., & Dressler, F. (2019). *Survey of event correlation techniques for attack detection in early warning systems* (Technical report 01/08). University of Erlangen, Germany.

Lippmann, R., Ingols, K., Scott, C., Piwowarski, K., Kratkiewicz, K., Artz, M., & Cunningham, R. (2007). *Validating and restoring defense in depth using attack graphs*. In: Proceedings of the IEEE military communications conference, Washington, DC, USA, October 2006.

Lockstep Consulting. (2004). *A guide for government agencies calculating return on security investment. Version 2.0.* Sydney: New South Wales Department of Commerce Government Chief Information Office.

LogRhythm SIEM. In: Logrhythm Off. website. https://logrhythm.com/. Accessed 27 Mar 2019.

Manadhata, P. K., Kaynar, D. K., & Wing, J. M. (2007). *A formal model for a system's attack surface* (Technical report CMU-CS-07-144). Carnegie Mellon University, Pittsburgh.

MAnagement of Security information and events in Service InFrastructures (MASSIF). In: Res. Proj. Eur. Community Seventh Framew. Progr. (FP7). Contract No. 257475. https://cordis.europa.eu/project/rcn/95310/factsheet/en. Accessed 27 Mar 2019.

Mell, P., Scarfone, K., & Romanosky, S. (2007). *A complete guide to the Common Vulnerability Scoring System Version 2.0.* FIRST Forum Incid Response Secur Teams, NIST.

Motahari-Nezhad, H. R., Saint-Paul, R., Casati, F., & Benatallah, B. (2011). Event correlation for process discovery from web service interaction logs. *VLDB Journal, 20*, 417–444. https://doi.org/10.1007/s00778-010-0203-9.

Müller, A. (2009). *Event correlation engine*. Master's thesis, ETH.

NMap. *NMap reference guide*. http://nmap.org/book/man.html. Accessed 27 Mar 2019.

Noel, S., Jajodia, S., O'Berry, B., & Al, E. (2003). *Efficient minimum-cost network hardening via exploit dependency graphs*. In: Proceedings of the 19th annual computer security applications conference, 2003, pp. 86–95.

NVD. (2019a). *National vulnerability database*. https://nvd.nist.gov/. Accessed 27 Mar 2019.

NVD. (2019b). *Common Platform Enumeration (CPE) official website*. In: NIST website. https://nvd.nist.gov/products/cpe. Accessed 27 Mar 2019.

Peltier, T. R. (2010). *Information security risk analysis* (3rd ed.). Boca Raton: CRC Press.

Poolsappasit, N., Dewri, R., & Ray, I. (2012). Dynamic security risk management using Bayesian attack graphs. *IEEE Transactions on Dependable and Secure Computing, 9*, 61. https://doi.org/10.1109/TDSC.2011.34.

QRadar SIEM. In: IBM Off. website. https://www.ibm.com/support/knowledgecenter/en/SS42VS_7.3.2/com.ibm.qradar.doc/c_qradar_oview.html. Accessed 27 Mar 2019.

Radack, S., & Kuhn, R. (2011). Managing security: The security content automation protocol. *IT Professional, 13*, 9–11. https://doi.org/10.1109/MITP.2011.11.

Sadoddin, R., & Ghorbani, A. (2006). *Alert correlation survey: Framework and techniques*. In: Proceedings of the international conference on privacy, security and trust: Bridge the gap between PST Technologies and Business Services.

Splunk Enterprise Security. In: Splunk Off. website. https://www.splunk.com/en_us/software/enterprise-security.html. Accessed 27 Mar 2019.

SPSS Tutorials: Pearson Correlation. In: Kent State University Libraries. https://libguides.library.kent.edu/SPSS/PearsonCorr. Accessed 28 Mar 2019.

Tenable. *Nessus vulnerability scanner*. http://www.tenable.com/products/nessus-vulnerability-scanner. Accessed 2 Jul 2018.

Tiffany, M. (2002). *A survey of event correlation techniques and related topics*. http://citeseerx.ist.psu.edu/viewdoc/summary?doi=10.1.1.19.5339. Accessed 30 Oct 2019.

Toth, T., & Kruegel, C. (2002). *Evaluating the impact of automated intrusion response mechanisms*. In: Proceedings of the IEEE annual computer security applications conference, Las Vegas, NV, USA, December 2002.

Tuchs, K. D., & Jobmann, K. (2001). *Intelligent search for correlated alarm events in databases*. In: Proceedings of the 2001 7th IEEE/IFIP international symposium on integrated network management, Seattle, WA, USA, May 2001. Integrated Management Strategies for the New Millennium, IEEE.

Wang, L., Jajjodia, S., Singhal, A., & Al, E. (2014). K-zero day safety: A network security metric for measuring the risk of unknown vulnerabilities. *IEEE Transactions on Dependable and Secure Computing, 11*, 30–44. https://doi.org/10.1109/TDSC.2013.24.

Wayne, W. D. (1990a). Spearman rank correlation coefficient. In *Applied nonparametric statistics* (2nd ed., pp. 358–365). Boston: PWS-Kent.

Wayne, W. D. (1990b). Kendall's tau. In *Applied nonparametric statistics* (2nd ed., pp. 365–377). Boston: PWS-Kent.

Wei, W., Chen, F., Xia, Y., & Jin, G. (2013). A rank correlation based detection against distributed reflection DoS attacks. *IEEE Communications Letters, 17*, 173–175.

Wireshark. *Wireshark vulnerability scanner*. https://www.wireshark.org. Accessed 27 Mar 2019.

Wu, Y. S., Foo, B., Mao, Y. C., Bagchi, S., & Spafford, E. H. (2007). Automated adaptive intrusion containment in systems of interacting services. *Computer Networks, 51*, 1334–1360. https://doi.org/10.1016/j.comnet.2006.09.006.

Xu, D., & Ning, P. (2008). Correlation analysis of intrusion alerts. *Advanced Information Security, 38*, 65–92. https://doi.org/10.1007/978-0-387-77265-3_4.

Zurutuza, U., & Uribeetxeberria, R. (2004). *Intrusion detection alarm correlation: A survey*. In: Proceedings of the IADAT international conference on telecommunications and computer networks, pp. 1–3.

Cybersecurity Technologies for the Internet of Medical Wearable Devices (IoMWD)

Raluca Maria Aileni, George Suciu, Carlos Alberto Valderrama Sukuyama, Sever Pasca, and Rajagopal Maheswar

1 Introduction

The field of *wearable* frameworks, including small-scale sensors entirely integrated into textiles, digital watches, belt-worn personal computers (PCs) with a HMD (head-mounted display), and glasses, in short, objects that are worn on different parts of the body, is intended for broadband activity. The domain of wearable health monitoring systems is advancing toward limiting the dimension of wearable devices, estimating vital signs, and sending secure and reliable information through smartphone innovation (Haghi et al. 2017).

Even though there has been an interest in observing the bio-/nonmedical information for monitoring the environmental, wellness, and medicinal data, one obvious use of wearable systems is keeping under observation physiological parameters in the mobile environment. Most of the economically accessible devices are used to monitor vital signs, but some of them are not appropriate for medical monitoring of high-risk patients. The devices used in the medical scope are usually straightforward (Indiegogo 2019).

Nowadays, people, especially the working class, spend most of the day commuting between different tasks, inclining to disregard their well-being and wellness (Xu et al. 2014). Because of that, even a simple meeting with a specialist in a clinic may require a few tests for diagnosis, prescription, and treatment, which can take much

R. M. Aileni (✉) · G. Suciu · S. Pasca
Politehnica University of Bucharest, Faculty of Electronics, Telecommunication and Information Technology, Bucharest, Romania

C. A. Valderrama Sukuyama
Mons University, Faculty of Engineering, Department of Electronics and Microelectronics, Mons, Belgium

R. Maheswar
VIT Bhopal University, SEEE, Bhopal, Madhya Pradesh, India

S. K. Shandilya et al. (eds.), *Advances in Cyber Security Analytics and Decision Systems*, EAI/Springer Innovations in Communication and Computing,
https://doi.org/10.1007/978-3-030-19353-9_6

time. Most of the patients would go to a clinic when they are facing severe illness. People are looking for an alternative option, for example, a device that can be worn on the body, which would continuously monitor the users' health in real time but also would supply different well-being parameters, just as the doctor itself.

The already existing hard security and privacy conditions of the healthcare domain are expected to exacerbate in this context due to the monitoring, collection, storage, sharing, and recovery of patient information but also the cooperation between different associations, systems, and health professional personnel (Burleson et al. 2012). A system of remote sensors can be networked to monitor somebody's physical or environmental specific features. This type of monitoring can help in recognizing the predisposition of particular illnesses for a particular population, area, or environment. It is also fundamental for early interventions, for example, alerting in case of any signs of specific diseases.

In the last years, two categories of well-being monitoring devices have been deployed: in-body devices (implanted) and on-body devices (wearable). The invasive devices such as pacemakers, defibrillators, and neuro-stimulators screen and treat physiological conditions inside the body. These devices have wireless capabilities that empower remote monitoring of the patient's health condition through an implant reader which can send data to the caregiver through the Internet. On-body devices, or wearable devices, consist of motion sensors and blood pressure meters and are frequently used to monitor the general health status of patients (Anwar et al. 2015).

At present, RFID (radio-frequency identification) technology is regularly utilized to monitor medical and human resources where there is a need to give prompt care. For instance, when patients search for emergency medical attention, a health expert should know if the patient has implanted several devices. For example, a doctor who is treating a patient who has an implanted pacemaker while suffering from a heart attack can attach RFID labels to the implant for device identification. The RFID reader allows specialists to access data about the pacemaker by using an NFC (Near Field Communication) chip embedded in a mobile phone that can react as an RFID reader when the mobile phone is turned on. After patients' data is gathered, healthcare suppliers can keep it in a place that it can be accessed quickly and it can be accessed only by authorized staff.

Internet connectivity is omnipresent nowadays and has led to an entirely different pattern – the Internet of Things (IoT), which represents the idea of interconnecting physical objects to one another or the Internet to produce domain-specific knowledge through data analytics and visualization with cloud computing (Hiremath et al. 2014). Throughout the years, as we moved from basic web administration to social networks, to wearable web, the interest for interconnecting intelligent wearables has expanded.

Internet of Things (IoT) has been widely deployed with limited to no security features. Security uncertainties started to arise since massive amounts of data are being collected, stored, manipulated, merged, analyzed, and erased; security concerns have begun to break out. The development of new technologies guided by policies to address device security issues throughout the life cycle of the data is needed.

The design of medical devices (Koydemir and Ozcan 2018) should involve a high attention to the selection of wireless technology, e.g., IEEE 802.11, Bluetooth, wireless medical telemetry service, wireless connection stability, wireless protection, and performance frequency with minimum interference in the operation band of other nearby radio-frequency devices, taking in consideration not only the country in which the product is made but also other countries where it is going to be used.

Usually, the MIoT architecture is composed of three layers (Sun et al. 2018): the perception layer, the network layer, and the application layer. The fundamental function of the perception layer is to assemble healthcare information with different types of devices. The network layer, which is made of wired and wireless systems and middleware, operates and transfers the input gathered by the perception layer sustained by technological platforms. The application layer integrates the medical data sources to implement personalized medical services and comforts the definitive users' needs, according to the whole situation of the target community and the service demand.

IoMT (Internet of Medical Things) refers to all the applications and medical devices which are associated with healthcare IT systems through computer networks. IoMT utilizes IT services as its base for the assistance of the medical devices to interact with the tools. Examples of IoMT applications are health monitoring, mHealth applications, etc. These applications allow patient users to transmit medical information over their phones and Internet connectivity in real time. Patients suffering from diabetic, cardiovascular diseases are considered suitable for the use of remote mobile applications for monitoring their health condition and for sending their medical records to the doctors. Such devices are utilized in hospitals to provide support to the patients and to prevent doctor trips to the patient's home.

A protected database is required to ensure reliability and safety in medical or healthcare societies. Cybersecurity is an issue commonly encountered in the IoMT on the account of the expanding manipulation of medicinally associated devices in the healthcare industry and the absence of an appropriate security framework in the medical foundations and associations. This may conduct data losses, sharing individual information, and changing medical information or more critical data about drugs and doses but also hacking of MRI and X-ray machines in the hospitals. Cybersecurity is relevant for these medical devices and applications because the vulnerabilities can incidentally or intentionally kill somebody. Thereby, cybersecurity should be executed on high standards in order to guarantee the well-being of the patients and the association's notoriety (Dr. Hempel Digital Health Network 2019).

Motion tracking has a couple of significant applications in sports, medical, and different domains. Such applications incorporate fall risk validation, evaluating sports exercise, and studying people's habits. Wearable tracking devices have become popular in the last years because they can motivate the users day by day when they are exercising, and they gather all the data through a smartphone in one single application (Anliker et al. 2004).

Specifically, to precisely monitor the movement of the body, three-axis accelerometers, magnetometers, and gyroscope sensors acquire information, each of them for particular reasons. These sensors can be utilized for human activity recognition in the computing area as well (Anon 2019). Additional sensors, such as gyroscopes and magnetometers, can be associated with accelerometers to compensate for the absence of accuracy in collected data for motion tracking. Usually, the mixture of these three sensors can generate to 9DoF (degree of freedom), a three-axis accelerometer combined with a three-axis gyroscope and a magnetometer (compass). Human movement detection has a wide range of use, from sports to entertainment or biomedicine. One of the main applications in motion tracking is healthcare. Because of the high-accuracy need in the biomedical domain, these devices are not usually appropriate for the medical application. In order to establish the notion of motion tracking in the medical application field, several clinical investigations have been carried out regarding posture estimation by using accelerometers (Mathie et al. 2003). The earlier form of the newest integrated sensors (accelerometer, gyroscope, and magnetometers), which supply high-accuracy data on 9DoF, was suggested in 2006, by Roetenberg (2006). Roetenberg proposed a set of triaxial accelerometers, triaxial gyroscopes, and a magnetometer for human motion estimation and monitorization.

Even various wearable devices have been designed to determine essential elements in healthcare monitoring, for example, measurement of the electrical signals from the heart (electrocardiogram) and brain (electroencephalogram) and the temperature of the skin estimation; a high attention should be on the data accuracy and security.

Numerous devices, structures, designs, and solutions for remote wearable ECG monitoring, which has an essential position in the field of health monitoring, have been proposed in the literature and industry. Typically, such solutions are trying to execute and are not sufficient enough when referring to power consumption. Several solutions are exceptional but do not have the likelihood of merging with other types of signals from different systems.

WBAS (wearable body area sensors) are front-end units of WIoT and discreetly cover the body in order to gather health-centric data. WBAS are in charge of:

1. Gathering data directly from the body through contact sensors or from peripheral sensors and providing indirect data of the body and its behaviors
2. Setting up the information for either accessible examination for closed-loop feedback or remote transmission for extensive investigation and choice help

WBAS, regardless of whether business or research facility models, are bundled with small dimension sensor hardware, an embedded processor with storage competence, power administration, and optional communication electric circuits relying upon the application. For example, external wearable sensors, such as the BodyMedia armband (Jawbone Inc., USA), are fitness monitors that operate on less computationally concentrated algorithms with least hardware requirements, their primary objective being to motivate people sustaining an active lifestyle. Almost all of the contact-type wearable sensors include a reasonable amount of electronics and computing abilities because they are desired to supply high-accuracy, high-resolution clinical data of patients in real time. A new interface between sensor and body is essential to adequate data acquisition in wearable applied science. Examples of such innovative sensor technologies that give constant access are chest-worn ECG monitor (Mankodiya et al. 2010), portable Bio-Patch (Yang et al. 2014), and ring sensors for pulse oximetry (Asada et al. 2003). Also, in the market, there are available wearable devices woven onto clothes for providing unnoticeable health monitoring for patients that live in their homes, far from medical clinics and specialists. Moreover, smart clothes with material-based sensors have proven to help check the independent nervous system response (Seoane et al. 2013).

As access to the World Wide Web, informally mentioned as the Internet, becomes omnipresent, it was inevitable that medical device producers would tackle this ability to make their devices easy to use, administrate, and maintain. Combined with the capacity to get close enough ongoing information from such devices, the cost has commonly been driven down, and utility has improved. However, cyber risks are increasing (Stanley and Coderre 2016).

The healthcare system has been afflicted for a long time by issues; for example, diagnosis writes down unreadably on paper and specialists not having the capacity to retrieve patient information without difficulty. With the recent developments in technology, there are many possibilities to improve the modern state of healthcare to decrease several issues and supply more customized service. For instance, the acceptance of the Internet by the healthcare associations in the last 10 years has given a middle way for distributing general health information, enabling patients to acquire a lot more comprehension about medical conditions (Meingast et al. 2006).

2 Internet of Medical Wearable Devices (IoMWD) Security Vulnerability Analysis

Healthcare systems are prone to security and privacy issues, provoking dangerous consequences to patients like denial of service. Some vulnerabilities might have a more significant impact on one specific component than another. For instance, secured communication will be more conspicuous in surveying and fast interventions, while issues related to application security are likely to be more prominent in self-care (Anwar et al. 2015).

The security and privacy issues will be presented over four distinct planes. The user plane contains issues related to the user's preferences, understanding, and competencies. Healthcare area includes various types of patients who perceive privacy in different ways, but usually, the security of health information might not be an urgent concern for someone. Patients would use any service that offers them health benefits. A study conducted in 2011 (Pew Research Center 2019) stated that 80% of Internet users are searching for health information online. Because of that, the phishing attacks aiming at users who search online for health information are inclined to success. Most patients do not comprehend the potential risks and consequences of utilizing health-related applications or instruments.

Within the application plane, there are included security issues of smartphone applications and third parties like online social networks (OSNs), which are on the rise because of the omnipresence of smartphones and the massive use of authentications via OSNs due to their rapid approach. The majority of health applications transmit data from user computers toward the application servers. Usually, users do not know what type of information is sent and neither who receives it. Mostly, the applications transmit the device's unique ID, the location, and, in some cases, even personal details. A clear privacy breach represents the "free" applications that contain ads and send users' location to the advertisers to target ads by location. There are cases in which applications were sent, as well as personal user information, to advertisers. Nowadays, the majority of applications are intended for smartphones, but their security systems are usually not proper. For instance, Android OS permits the specification of the policy that administers access to application interface or data while it is installed. Still, the application has a reduced capability of managing the rights among users. Therefore, a mobile health application might unintentionally permit a malware to access its interface or even data. For example, SMS phishing is when a malware pretends to be a "friendly" healthcare application that can access contacts, text messages, call history, and video.

The malware may utilize text-messaging APIs (application programming interface) to transmit false signals to the contact list or to block messages. Also, through the smartphone's voice-recording API, a malware application can record private conversations.

At the communication plane, several actions consist of devices interacting one with another, gathered data from remote devices being shared with caregivers, and caregivers remotely communicating with devices in order to perform specific commands. Patient data usually needs to be sent from the monitoring devices to a caregiver for offering treatment of quality and providing appropriate care. Preferably, a caregiver must transfer the collected data in a patient's health record to another caregiver. Generally, patient health data is usually sent through individuals and devices when delivering healthcare, and this facilitates security attacks and privacy breaches of sensitive health data. Sometimes, health data is self-reported, and other times, it is transmitted to healthcare suppliers by using different communication channels. On the way, the data stream can be noticed, which may result in the disclosure of a specific patient's treatment for a particular condition. A breach in location privacy is when following the communication from source to destination nodes; the locations of both patient and supplier can be found. The communication channels which connect nodes, for example, patient and supplier, can be designed utilizing different physical media or routing protocols. Physical layer security includes the security of media, signals, and binary transmission of information. The physical layer of systems must cope with various attack types including eavesdropping, jamming, and capture attempt. The majority of monitoring devices are intended to communicate with healthcare suppliers or health intervention frameworks remotely and automatically. For instance, defibrillators, pacemakers, and insulin pumps transmit wireless signals. Likewise, for some geriatric care cases, every user conveys more devices or sensors which are additionally associated with a similar system and interoperate among them. Therefore, issues of communication security within wireless systems can become frequent in a developing healthcare network. Because there are no physical boundaries within a wireless transmission, this may easily be retrieved by an attacker. One of the most popular wireless standards is Bluetooth technology, which is widely used within mobile health applications. This is subject to MitM (man-in-the-middle) attacks, device discovery attacks, and battery depletion attacks.

Referring to the device plane, security issues are related to storage and therapeutic or monitoring instruments. There is a large number of medical devices, which represents a reason for complications in the security prospects within the healthcare area of expertise. Having roughly $133 US billion in 2016 (Managed Care 2016), the US market for medical devices was the largest in the world. An exposable wearable device nearly always is on the risk of having privacy breaches. Occasionally, minor exposure to a device may not represent a concern; however, by distinguishing the device's type, particular conditions may be uncovered. A patient might not want to reveal certain circumstances, due to social disgrace. Because a device follows its patient, by tracking it, the patient's location becomes uncovered. Therefore, the simplicity of monitoring a device raises concerns about location privacy. Discussing on device security issues, they may be software-related or mechanical. The control of medical devices is provided by firmware software that have routines which run on privilege mode to translate and display sensor data in monitoring devices like blood pressure and heart rate, to allocate medication, to interpret raw information that will further generate human-understandable images like computed tomography (CT) scan, to offer therapy through defibrillation or cardiac pacing, or to act upon a doctor's intervention.

An attacker can exploit defects in the firmware, such as inconsistent error handling and buffer overflow. An attacker can also compromise confidentiality, integrity, or availability of information gathered by medical devices. By launching a confidentiality attack, an adversary gets illegitimate access to the devices and so to the patient's identification data. Through an integrity attack, the information enclosed by medical devices can be modified. In an availability attack like the sleep deprivation attack, the devices can become inaccessible to the authorized user. Information integrity and availability are mandatory in offering opportune healthcare, in both quality and time features. There exist inside attackers, who can be hardware/software professionals, clinicians, or even patients themselves. Eavesdropping on RFID readers represents a significant threat because RFID tags have been broadly utilized to recognize implantable devices. Subsequently, a patient with embedded medical devices can become a victim of stalking and other wrongdoing.

The most important categories of implantable medical device (IMD) vulnerabilities that have been identified are privacy vulnerabilities, where an IMD discloses a patient data to an unauthorized entity and control vulnerabilities, in which a hacker may take control of an IMD's operation or even disable its remedial services (Burleson et al. 2012). The two types of vulnerabilities might be harmful to patients' well-being results, and both are avoidable.

Because MIoT devices do not have enough capabilities when referring to computation, memory, and transmission, an effective and scalable high-performance computing (HPC) and substantial storage infrastructure for data storage and real-time processing are required. Nowadays, almost all MIoT institutions deploy their application servers and keep the gathered medical data in the cloud. Being flexible, cloud services facilitate efficient management of common healthcare data. Solutions for securing these data are enumerated as follows:

- Data Encryption. Cryptography represents a branch of security communication techniques in compliance with agreed rules (Zhao 2016). The plaintext, the initial message, is encrypted into ciphertext by an encryption algorithm; it is sent via a public channel to an intended receiver, and then, the message is decrypted back into plaintext. Data encryption can be developed at three transmission dimensions: node, link, and end-to-end encryption. In link encryption, for any transitional node, the message from the previous link will be decrypted into plaintext, and then, at that point, it will be transformed back into ciphertext utilizing the secret key of the following link. Still, node encryption does not permit messages in plaintext to build in the network node, like in the link encryption. Accordingly, node encryption may offer high security toward network data. When utilizing end-to-end encryption, the message remains encrypted until the recipient receives it. Since messages are constantly present as ciphertext all through the transmission, there is no exposure of data regardless of whether a node is corrupted or not. To protect eHealth transmissions, protocols for essential administration have an indispensable job in the security procedure. Nevertheless, complex encryption communication protocols or algorithms can significantly influence the transmission rate and even neglect to perform information transmission. Also, they have to possess valuable medical assets which are not accessible. The vigorous harmony between security assurance and system energy utilization should be understood with cautious and scientific steps.
- Access Control. Access control represents the method through which data systems specify the user and the predefined procedures that will prevent unauthorized users from retrieving specific resources (Bacis et al. 2016). In access control, there are different encryption methods, including symmetric and asymmetric key encryption (SKE and AKE) and attribute-based encryption (ABE). Cryptography depends on keys, the size, and generation method straightforwardly influencing the security of the cryptosystem. Thus, within a cryptosystem, a key administration technique decides the security framework's life cycle. Patient health data can be shared electronically with the specific authorization of data exchange in an auditable way, within the Health Information Exchange (HIE). Nonetheless, existing methodologies for permission in healthcare data frameworks show a few disadvantages in addressing the necessities of HIE, with non-cryptographic methodologies not being entirely reliable and secure systems for accessing policy enforcement, while cryptographic methodologies being excessively costly, complex, and restricted in indicating policies.
- Trusted Third Party (TTP) Auditing. Cloud servers are not entirely trusted. If information corruption or even erasing occurs without the client's consent, the consistency and integrity of medical information stored in the cloud can become compromised. Because of security issues, the rules are regularly stated by the

client, so the server supplier does not straightly have contact with the source information. Also, the trusted third party (TTP) having an excellent reputation that offers the impartial examination results can be appropriately introduced, to empower the responsibility of the cloud service suppliers and ensure the legitimate advantages of the cloud clients. TTP has some research issues which comprise dynamic and batch auditing. Over the previous decades, numerous auditing strategies have been introduced. Some machine learning (ML) methods, like support vector machine or logic regression, have been utilized to distinguish suspicious access. At the moment, unsupervised methodologies progressively draw more awareness.

- Data Search. For securing the privacy, sensitive information must be encrypted before deploying it, which obsoletes conventional information usage dependent on a plaintext keyword search. Therefore, it is of great importance to enable an encrypted cloud service for data search. Searchable symmetric encryption (SSE) and public key encryption with keyword search (PEKS) represent important strategies for searchable encryption. Furthermore, it is noticeable that the more complex the encryption measures are, the more difficult the information is sought, and the more complicated the consistency of search items is verified. All privacy and security measures have less significance if the items that will be searched cannot be enforced on time.
- Data Anonymization. The sensitive information of patients can be split into three classifications: privacy properties, explicit identifiers, and quasi-identifiers. Privacy data concerns the sensitive attribute of a patient, including illness and salary. Explicit identifiers can particularly indicate a patient, for example, through name, an ID number, or a mobile phone number. A mix of quasi-identifiers can likewise determine a patient, for instance, through the address or birth date. The datasets must be appropriately processed to secure the patient's data.

Referring to security vulnerability issues within medical wearable devices, Fitbit devices will be analyzed (Cyr et al. 2014). Fitbit devices hit the market in 2007 and are aimed toward a healthier lifestyle. As Eric and James, the co-founders of Fitbit, saw the potential wireless wearable devices would bring to society, they decided to develop the wearable fitness smartwatches Fitbit.

Wearable devices are worn daily by many people, but they do not have a clue on what threats their privacy can suffer by sharing their personal health information toward a company or a third party such as a cloud server. Various studies were made, and it was proven that Fitbit devices (Seals 2019) (Fitbit One and Fitbit Flex wristbands) have a vulnerability, whereas researchers at University of Edinburgh discovered that there is a way of intercepting data which is transmitted between a Fitbit device and a cloud server. They could access private data of Fitbit users and create false activity records, allowing them to send it to third parties which are unauthorized to know that data. The impact of knowing what health issues some people have could affect their well-being and intimacy. The answer to this is that some third-party people, by identifying such delicate information about a person, could not allow the specific person not to be hired at a particular company or even threatened with it for various reasons.

3 Technologies and Architectures for Secure IoMWD

Cybersecurity is related to confidentiality, integrity, and availability (CIA) triad. The comprehension of cybersecurity regarding CIA triad can facilitate thinking and decision-making. Confidentiality keeps sensitive information from not being seen or retrieved by unauthorized individuals while checking that those that have authorized access need to get to the information can access that information. As an example, data encryption can encode medical information such that only owners of a decryption key can read it. Encryption represents one of the principal methods used to secure patient information. Integrity implies guaranteeing that information continues to be precise and constant for its life cycle. For instance, it is essential to secure the data integrity of medical outcomes, keeping hackers away from making alterations in a diagnosing, from a positive to a negative, or perhaps modifying blood classification or hypersensitivity information. Availability points out the significance of keeping PC frameworks on the web and available when they are mandatory by the business organization. Denial-of-service (DoS) assaults represent a usual hacking procedure that is used to overburden computer supplies for not being accessible to authorized users (Stanley and Coderre 2016).

Almost all medical devices will fall into one of these models, and in order to have better comprehension, a specific technical procedure or control has been performed. It has to be noticed the fact that, while medical healthcare systems are mostly concentrated on confidentiality, integrity is perhaps one of the most critical dwellers. Risk management process must be considered prior despite all of the security issues.

Figure 1 illustrates a study performed on five clusters of data using LDA (Latent Dirichlet Allocation) modeling algorithm which considers subjects such as privacy and security, wireless network technologies, applications (e.g., design or development of device/application in healthcare, e.g., patient monitoring systems), data, and smart health and cloud (Dantu et al. 2019).

Healthcare represents the main topic when discussing HIPAA (Health Insurance Portability and Accountability Act) adjustments, which gives information security and safety dispositions for defending medical data. IoT assimilation in the medical field can be supported with improved IoT safety and privacy characteristics. A series of studies have proposed several IoT frameworks being made up of regulations, instructions, protocols, and standards that empower the implementation of IoT applications in different fields (Ammar et al. 2018; Lin et al. 2017). Considering the requirements for ensuring protection and confidentiality of patient medical data and also assuring electronic medical records and features, such as privacy and security, in IoT field are representing a significant aspect to researchers and experts.

A case study, elaborated in the Methodist Hospital of Southern California, was performed by the biomedical engineering team and has perceived the need to oversee medical devices by describing the vulnerabilities related to every device. In order to achieve this, the team is building up a program that reconsiders medical equipment administration. Additionally, for improving the workflow and data administration activities, this ascension in technology has acquainted new vulnerabilities regarding patient safety, data accessibility, and cybersecurity.

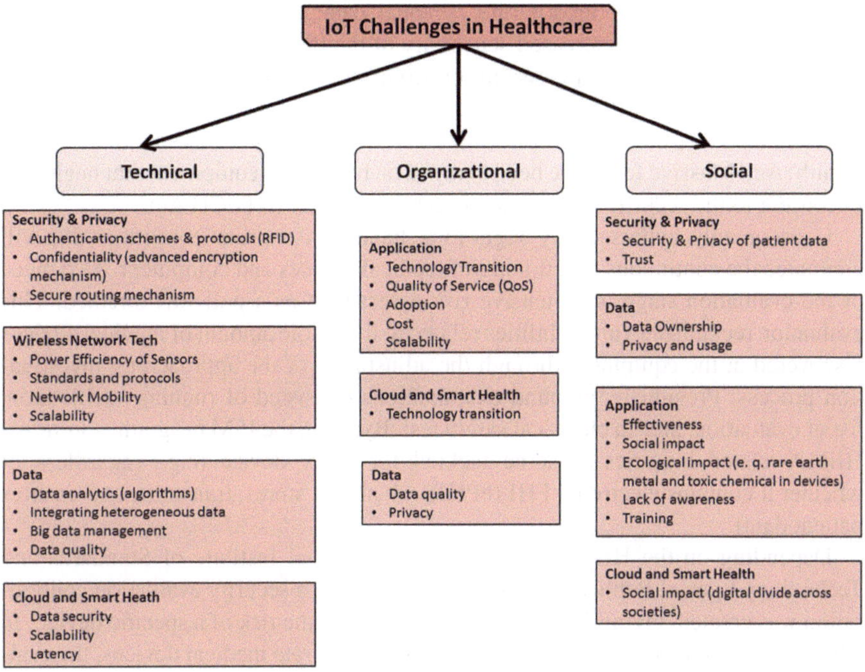

Fig. 1 IoT challenges in healthcare

The Methodist Hospital Biomedical Engineering Department, with the support of Renovo Solutions LLC, predicted the requirements to tackle the risk related with the development in technology and, as a result, performed an ISM (Integrated System Management) program. The ISM program comprises three stages: (1) risk assessment, (2) mitigation, and (3) recurrent management.

Nonetheless, the interoperability and interconnectivity between medical devices bring out cybersecurity concerns that were already known in the healthcare business. Before, medical devices were not associated with the hospital's networks. Even if they were vulnerable to infections through non-network circumstances, cybersecurity risks were fewer in contrast with the devices included in existing networks. Because there are a lot of medical devices connected to the hospital's network, they reveal several issues which were not present in older devices. Devices that depend on off-the-shelf software, more explicit, operating systems (e.g., different versions of Microsoft Windows) are powerless against various types of attacks, for example, malware or viruses.

Medical devices can be threatened using different approaches, and the primary ways by which cybersecurity threats can injure healthcare equipment are by obstructing (1) functioning of medical devices and (2) the integrity of the data. Specialists have clearly shown that the functioning of individual medical devices can be obstructed when they are not giving appropriate patient consideration. Due to the high importance of medical devices, cybersecurity threats can affect patient safety. Medical devices and IT systems may also contain PHI (protected health information) which can incorporate personal data, medical documentation, and payment data. With the increased sensitivity of the PHI, hospitals are the main targets for spyware and phishing assault for acquiring specific information. Utilization of PHI is controlled by the HIPAA (Health Insurance Portability and Accountability Act) and the HITECH (Health Information Technology for Economic and Clinical Health) Act. Massive fees have been charged on healthcare equipment that neglected to consent to the security and privacy necessities of the HITECH Act.

Consequently, cybersecurity ought to be handled as a prime concern as this will guarantee the continuous functioning of medical devices and equipment. As a piece of the evaluation stage, an intensive risk assessment as a poll was directed. This evaluation recognized vulnerabilities related to the whole amount of medical devices discovered at the equipment through the adjustment of the approaching investigation process. Previously, inbound examinations consisted of running verification, list of evaluation, and an electrical safety test. By using the ISM program, additional HIPAA-related inquiries must be tackled for every device (e.g., regardless of whether it contains electronic PHI (ePHI), whether it stores, transmits, and ensures patient data).

Depending on the HIPAA guidelines and National Institute of Standards and Technology Special Publications 800-30 and 800-66, a security evaluation configuration was created, giving 57 inquiries for evaluating the risk of a specific device. In addition to the data utilized by the systems to incorporate medical devices, the evaluation concentrated on commands, strategies, and action plans that influenced the confidentiality, integrity, and availability of ePHI. The risk assessments are measured and assembled into three classifications: confidentiality, integrity, and availability. The individual groups are then merged and systematized on a scale from 0 (lowest risk) to 100 (highest risk). After that, the evaluation grants for restorative proceedings are to be conducted in order to diminish the recognized risks. The manufacturer then requests the evaluation and MDS2 structure.

At Methodist Hospital, the engineering group decided to build up a mitigation plan for every medical system recognized in the risk assessment, and as a consequence, the standard preventive support through the life cycle of a medical device has been reevaluated to consolidate application of the ISM program, such as checking virus protection, supplying merchant-endorsed patch management, hardware administration (for servers and working stations), disaster recuperation (backups, hard drive ghosting), data security, and implementing strategies and plans of actions. The security assessment uncovers general vulnerabilities over the networked medical devices, such as essential administration, logging checking and examining, backups, support for the operating systems, and virus protection, which are tackled through the mitigation stage and kept up in the planned maintenance. After a better understanding of the actual risks and act by the proposed restorative activities distinguished by the ISM program, Methodist Hospital has turned out to be better straightened with the most recent practices in the industry and prepared for new vulnerabilities introduced by the headway of medical device technology. The IT security strategies and techniques adopted by the hospital alongside the ISM program have created a balanced cybersecurity program that assists the sustenance of healthcare.

Strengthening resilience represents the primary objective of the cybersecurity. Resilient associations are less likely to have their security broken and endure fewer damages when an infiltration happens (Martin et al. 2017). An effective way by which resilience can be enhanced is by keeping secure and up-to-date backups in order not to lose information permanently after an attack. In the event of a cybersecurity attack on Papworth Hospital in 2016, a ransomware infection accidentally happened soon after the daily backup, and no information was lost (Muncaster 2016). Other technique for upgrading resilience is insurance, a quickly developing business with worldwide offers of $2.75bn in 2015 (PriceWaterhouseCoopers 2016). The increasing expenses may cause insurance agencies to step with attention in the future. Medical service suppliers need to discover practical approaches to secure themselves against the possibly devastating expenses of cyberattacks, similarly as they do both with the expenses of clinical carelessness. Cybersecurity can be additionally reinforced by public help for incident management, hierarchical readiness, and risk guidance. The mechanism for giving such help is starting to develop, for instance, CareCERT, available in the UK (Mansfield-Devine 2016).

Another study (Lounis et al. 2016) depicts the design of an architecture which empowers a healthcare organization, for example, hospital or a medical center, to control data flow gathered by WSN (wireless sensor network) for patient supervision. The proposed design architecture (Fig. 2) can be expanded, and it is capable of storing a large amount of information achieved by the sensors. Due to the increased sensitivity of the data, a new security system is used to ensure information privacy, integrity, and fine-grained access control. In the proposed solution, security configuration and key management are straightforward to clients (patients and specialists) and do not demand their mediation. The architecture has two classifications of users: patients and medical service experts, and it consists of the following elements: (1) the WSN, which collects information from clients, (2) the monitoring applications which permit healthcare specialists to access the gathered information, (3) the HA (Healthcare Authority) which indicates and authorizes the security approaches of the social insurance foundation, and (4) the cloud servers which guarantee information stockpiling. By collecting data in the cloud, the design offers endless capacity limit and scalability. The design expands its stockpiling limit, through on-request provisioning highlight of the cloud. Also, it offers increased accessibility to medical organizations because it is not affected by the unpredictability of the server's administration.

4 IoT Edge and Cloud Cyberattacks

In the last years, technology processes have caused an extreme need for devices that have high computational power. Nowadays, there were significant changes that played an essential role in the way things evolved.

The health industry is probably one of the most important fields where IoT applications can be used at their maximum potential.

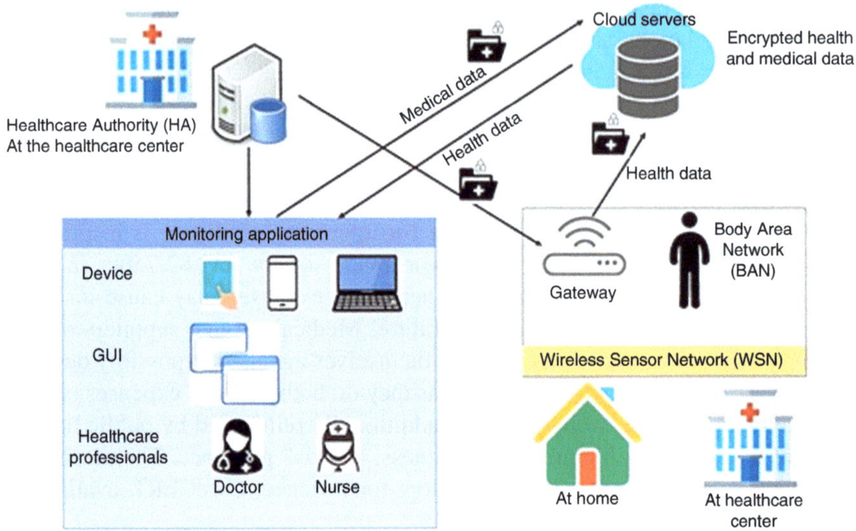

Fig. 2 Design architecture

Biomedical wearable devices gained a significant role considering the lifestyle that people have nowadays as they spend many hours between different tasks during the day.

Therefore, they usually do not have time to go to doctors, and consequently, the process of having the data sent directly to a physician can be an essential improvement. This way, doctors can offer medical suggestions based on the data that were taken from the patient. Finally, data can be stored on cloud servers and used later (Qureshi and Krishnan 2018).

In general, IoT has an architecture based on four levels: sensing, networking, service, and interface, as shown in Fig. 3 (Farahani et al. 2018). The role of the sensing layer is to connect to the real world through different hardware devices to collect data. The network layer ensures the data transfer via wired and wireless networks while offering also network support. The service layer has the role of creating and managing different types of services to meet the needs of users. The interface layer is the user interface that provides ways to interact with users, as well as other applications to analyze collected data.

Considering the close link between healthcare and technology, the rapid evolution of IoT and the acceptance by the general public of small wearable biosensors have created new opportunities for personalized eHealth and services. Thus, the four-level architecture mentioned above can be applied in the field of smart healthcare services and smart wearable devices as follows. At the sensing level, smart healthcare services can collect different types of data through sensors on cyber-physical or wearable systems. Then, at the network layer, using wireless networks, all data can be transported to servers or cloud. The service layer provides the healthcare service such as patient data processing regarding heartbeat, blood pressure, and

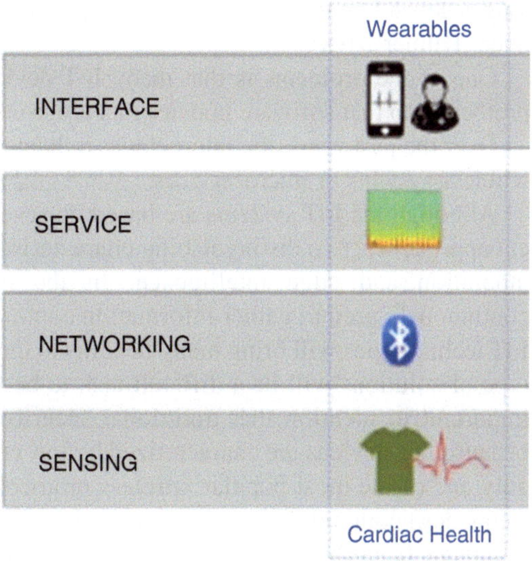

Fig. 3 IoT layers

glucose level. The interface layer provides an interface which allows users to retrieve and understand the collected data. Both patients and doctors have access to this interface to enable physicians to make appointments and patients to find details of their health without medical checkups. Security is essential in IoT eHealth as maintaining a high level of security in this area remains a challenge. For each device to be identified and tracked by bar code and intelligent sensors, RFID (radio-frequency identification) and WSN (wireless sensor network) technologies can be used.

Moreover, to ensure secure network security, network authentication and network firewall can be provided. IoT eHealth services can provide authentication, authorization, confidentiality, and integrity for all data and services. To maximize the level of protection, all the passwords will be authenticated and checked through the bar code on the hardware and personal keys on the software. Also, to minimize the possibility of leakage of personal information, at the top layer level, each user identity and location privacy can be kept anonymous.

The demand for a network that can enhance the possibility of sending a high amount of data has become an important issue nowadays. However, this aspect brings new challenges in terms of cybersecurity. For example, in the IoT filed, the loss of data occurs in different scenarios. Many devices can control and initiate cyberattacks, regardless of their characteristics (Pan and Yang 2018; Endler et al. 2017).

The increasing number of IoT devices that have low resources can cause security breaches that will lead to severe economic problems, especially for those that are physically remote. Therefore, edge computing will bring a new network infrastructure that will analyze and process data at the edge of it, rather than transport it to another remote data center. If an action is needed, the central server will send the response to the device after receiving and analyzing the data. The IoT device will no longer depend on the Internet connection and will work as an independent network node (Palmer 2018).

One of the problems is that many IoT devices do not come with IT hardware protocols, so that software updates that are usually required may not be available.

Over the past years, the Internet was influenced by different technologies such as wireless systems or microservices.

Although the IoT systems are based on several types of architectures or different approaches, two distinguishing characteristics are common to all of them: collaboration and edge intelligence. In the future, wearable devices will be continuously used to gather information coming from patients. Even though using IoT technologies will bring many benefits to the healthcare systems, providing this type of solutions will be a difficult task to be accomplished. In this context, it is important to mention that traditional security measurements cannot be applied because IoT devices are characterized by low computing power. Bluetooth is probably one of the most popular wireless protocols. It is mainly used for RF (radio-frequency) communications such as mobile phones, laptops, or wearable devices. Although from the first version Bluetooth improved its security methods, it still presents a couple of vulnerabilities such as Bluejacking or Bluesnarfing (Langone et al. 2017).

Several cloud computing vulnerabilities might raise security problems among users. Usually, the information that is stored in the cloud is sensitive and can be easily lost or destroyed as the cloud service provider cannot predict each type of threat that might occur in time. Therefore, encryption algorithms can be applied to sustain the integrity of data. Other significant vulnerabilities are related to APIs (application programming interface) as these can interfere in processes such as the management of the cloud service. By exploring the weaknesses that are associated with any cloud system, hackers can get unlimited access to the host.

Attacks that are intended to affect a cloud system have an impact on the communication between the users and services. They can get access to users' credentials and use legitimate access later.

Nowadays, there are several different ways to attack cloud services, and by being aware of some of them, cloud developers might come up with more secure solutions.

Attacks based on cloud malware injection rely on an infected module that is added to a SaaS (Software as a Service) or PaaS (Platform as a Service) solution. The most common forms of this kind of attacks are scripting and SQL ones. DoS attacks are performed to make some services unavailable to its users. Due to high workload, cloud systems tend to use multiple virtual machines, and users that have legal access might not be able to use the cloud as it will work too slow. Side channel attacks are implemented using a malicious virtual machine near the target virtual machine. Hackers use a system made of several cryptographic algorithms. The wrapping attack is a threat to cloud computing as users connect to this kind of service using a web browser. Typically, XML signature protects the credentials, although it still allows the attackers to move the document from one place to another (Bryk 2018).

However, even if IoT represents a vulnerable point for cyberattacks, edge networking can help improve data privacy. Although the information is shared between the devices that produce it, it is hard to compromise the entire network using a single attack.

Another advantage of using this technology is that, by storing and processing the data close to its source, the delay is significantly reduced. Therefore, the analysis is performed in real time. Furthermore, there is no need for a large amount of storage in the cloud, as the useless information will be removed once the data is saved "at the edge." This feature will also lead to lower infrastructure costs and expensive business operations. Since the device will work without being connected to the Internet, the problems related to the connection itself will not be so influential anymore. Unlike the cloud-based systems, edge computing can be scaled customized according to the characteristics that are required (Aleksandrova 2019).

5 IoMWD Cybersecurity Framework

According to Cisco definition, the cybersecurity involves protection of the IoT systems, servers, mobile devices, networks, software apps, and data from digital attacks. The threats countered by cybersecurity, according to Kaspersky, are:

1. Cybercrime (actors targeting systems to cause disruption)
2. Cyberattack (politically motivated information gathering)
3. Cyber-terror (actors targeting electronic systems to cause panic or fear)

The methods used to control IoT devices or networks include viruses, worms, spyware, Trojans, and ransomware.

ENISA (European Union Agency for Network and Information Security) defines IoT as "a cyber-physical ecosystem of interconnected sensors and actuators, which enable intelligent decision making" (ENISA 2017).

According to ISO/IEC 27032:2012, cybersecurity means "preservation of confidentiality, integrity and availability of information in the Cyberspace" (Brookson et al. 2016).

According to X.1205 ITU-T, cybersecurity means "collection of tools, policies, security concepts, security safeguards, guidelines, risk management approaches, actions, training, best practices, assurance and technologies that can be used to protect the cyber environment and organization and user's assets" (Brookson et al. 2016).

According to special publication 800-39 NIST, cybersecurity means "the ability to protect or defend the use of cyberspace from cyber-attacks" (Brookson et al. 2016).

According to CNSSI No. 4009, cybersecurity means "prevention of damage to, protection of, and restoration of computers, electronic communications systems, electronic communications services, wire communication, and electronic communication, including information contained therein, to ensure its availability, integrity, authentication, confidentiality, and nonrepudiation" (Brookson et al. 2016).

Concerning the cyberattack, according to ENISA study (ENISA 2017), the critical attack scenarios are envisaged to compromise IoT administration system (88.89%), to distort the sensor readings by replacing with other false values (84.72%), to compromise the hardware of the system by injecting commands (81.94%), to compromise the communication (73.61%), and to distort the information in cloud or sent to aggregators (73.61%).

Sensors and actuators are small system components, connected or not to the cloud through gateways, that can be integrated into wearable IoT devices. In healthcare area, several wearable IoT devices are embedded system, such as medical implants based on low-power consumption (ENISA 2017). However, it is already known that low-power consumption devices come with risks of cyberattacks, and it is estimated that 25% of cyberattacks will be targeting IoT wearable devices. IoT wearable device is having BLE (Bluetooth Low Energy) mesh network secured by AES-CCM cipher with 128-bit key length that prevents some security breaches. In order to reduce the security risks on IoT devices, MIT proposed another technique called elliptic-curve encryption.

BLE devices communicate with each other using (Boesen and Dragicevic 2018):

- Point-to-point network topology (piconet) and allowing one-to-one device communications
- Mesh network having a many-to-many topology, which means that all device are able to communicate with every other device in the mesh

Medical cyber-physical systems (MCPS) represent the critical healthcare integration of a network with medical devices. These systems are being used progressively in hospitals to get some high-quality healthcare services. MCPS are facing numerous challenges, including inoperability, security/confidentiality, and high security of software systems. Infrastructure on cyber-physical systems has been added to high-tech medical systems to increase the efficiency and safety of healthcare. It can also help doctors overcome critical device and challenge issues with which the medical device network faces. The concept of social networks and security, along with the idea of wireless sensor networks, is also introduced (Dey et al. 2018).

Healthcare system requires a constant improvisation in its organization resources and structure. Accordingly, many health research organizations manage to improve the efficiency and reliability of electronic health records (EHR). The medical institutions improved their proficiency through unification adapters and health monitoring devices over the network module. These organizations also make an operable function over the susceptible variables cached in their healthcare server. However, the operations defined in the server defect in their vital extensions as the structure of the healthcare system are more complicated than the predicted one. The modifications that happen frequently or rarely in the server frameworks can affect the service delivered by the wellness program. The changes can affect the service standards by performing an unusual behavior. For example, a doctor or medical assistants will be unable to provide proper treatment to patients in a given time due to irregular update

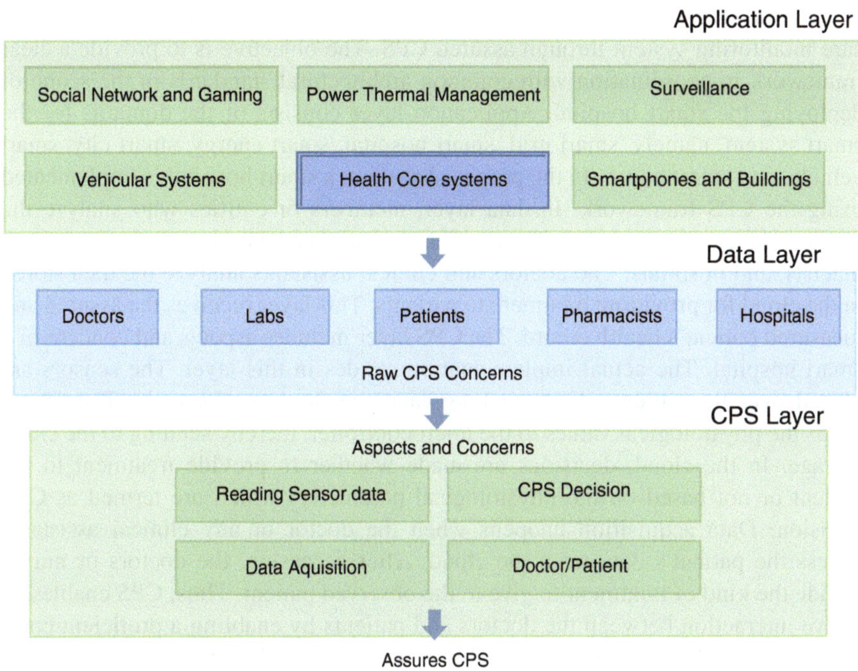

Fig. 4 A smart system based on cyber-physical system

along with unexpected costs. Hence, a smart system is required by integrating the service-oriented cloud with other smart solutions to monitor the patients regularly. The patient's health parameters are observed by sensors, microcontrollers, and other smart devices such as computers and mobiles. The interconnected solutions are accessible to clinical data which is presented through some algorithms and frameworks. The patterns are recognized through the algorithms for each patient with the responses stored in the data servers (Monisha and Babu 2018).

The smart system with effective machine-to-machine communication is provided through a cyber-physical system (CPS). CPS framework (Fig. 4), deployed for an efficient healthcare monitoring system, is a mechanism developed using problem-solving algorithms connected to Internet users through network adapters. CPS is a technique built upon logically by merging the optimized algorithms with the networks and smart, physical devices. CPS is employed in the platform whenever a smart implementation is required in an environmental application. In Fig. 4, a framework is designed for healthcare monitoring system by applying CPS notions. The framework structure is divided into three layers, namely, (a) application layer, which consists of the applications defined for CPS technologies; (b) data layer, which includes entities or the members who analyze data for further concern in the system; and (c) CPS layer, a layer consists of actual CPS implementation for smart hospital.

Each layer in the defined framework makes a vital scope for effective healthcare monitoring system through assured CPS. The objective is to provide a clear framework in coordination with common architectural standards in the scope of deploying the smart hospital. Application layer consists of the domains for the smart system, namely, smart grid, smart hospital, smart energy, smart city, smart vehicle, and smart house. In the proposed system, a smart hospital is implemented using the CPS framework. In data layer, members or entities who analyze the medical data are represented. The entities are patients, laboratories, doctors, pharmacists, and hospitals. The doctors and clinical assistants analyze the data stored in the cloud for providing treatment to patients. This layer receives the assured and measured patient's health record. The CPS layer includes aspects and concern of a smart hospital. The actual implementation resides in this layer. The sensors are placed over the patient's body making each sensor area as a node. The sensor sends the physiological values to the microcontroller, thereby sending to the cloud storage. In the cloud, decisions are made whether to provide treatment to the patient or not based on the physiological parameters which are termed as CPS decision. Data acquisition happens when the doctor or any clinical assistants access the patient's data from the cloud. After accessing, the doctors or nurses decide the kind of treatment to give to the observed patient. Thus, CPS enables an active interaction between the doctors and patients by enabling a proficient communication and computation model over the network. Hence, CPS provides an assured mechanism or algorithmic concept for implementing smart hospital (Monisha and Babu 2019).

An alternative insurance model that combines good engineering practice with information security is modified Parkerian Hexad for cyber-physical systems (operational technology) that added safety and resilience attributes, noting that availability includes reliability. The resulting insurance model presents the eight facets which address safety and security from three perspectives:

- Confidentiality involves the control of access and prevention of unauthorized access to systems and information or data.
- Integrity involves maintaining the consistency, coherency, and configuration of data and systems and preventing unauthorized changes to them.
- Authenticity ensures that inputs and outputs from systems, system status, and all associated one's processes, information, or data are authentic and have not been altered or modified.
- Utility ensures that the system and all information or data remain usable and useful throughout the product life cycle system and, where applicable, may be transferred to any successor system(s).
- Availability ensures that systems, information, or data and associated processes are consistently accessible and usable on time. To get the necessary availability, everyone can have an adequate and proportionate level of resilience.
- Control involves design, implementation, operation, and maintenance of systems and associated systems processes to prevent unauthorized control, manipulation, or interference.

- Resilience is the ability of systems and information or data to transform, renew, and recover in timely response to adverse events.
- Safety involves design, implementation, operation, and maintenance of systems and related processes to prevent creating dangerous conditions that can lead to injuries or loss of life or unintentional damage to the environment (Piggin 2017).

6 Future Challenges on IoMWD Security

The security and the confidentiality of data that are related to patients are two linked concepts. The first one ensures that data is securely transferred and stored, while the second one provides the possibility of accessing the data only by the people who have the authorization to use it (Haghi et al. 2017). Reasonable protection strategies can be used in different situations according to the requirements associated with each application. Medical wearable devices gained popularity in recent years as they provide useful information regarding people's health. However, this advantage comes along with additional problems in terms of information security and the protection of the collected data.

When developing the design of any system, the developer must take into consideration the impact of different factors and find the right balance between all of them.

One of the challenges in terms of security is set up by the network itself. Several devices or software rely on a wireless network which is known to be vulnerable to different intrusions like unauthorized access, attacks like man-in-the-middle, or traffic injections. Moreover, many wireless networks can be found in public and uncertified places.

Another challenge is represented by the policy and proxy rules applied to low-cost devices or applications. Therefore, high-grade security comes at higher costs. In the future, different levels of security protocols need to be developed for each user that will ensure security protection.

Another problem is represented by the fact that the standards associated with the collected data coming from different manufacturers can frequently change in time. Moreover, the patient's private data are sensitive and come along with many problems in terms of security.

Patients' sensitive data are collected from wearable medical devices using sensor networks which are more sensitive to cyberattacks than fiber networks. These types of cybersecurity threats can be classified into two main categories: passive and active. The first category refers to the possibility of changing the network routing configuration. The active attacks refer to more *critical* facts like resonate transmission of a patient data, alteration of medical device configuration in order to create harmful situations for the patient's health, eavesdropping on medical data, and malicious usage (Al Ameen and Kwak 2011).

The most common methods against cyberattacks are data encryption and authentication mechanisms. Due to the sensitivity of personal data, not mentioning their medical characteristics, encryption is necessary.

Acknowledgments This work has been supported in part by UEFISCDI Romania and MCI through projects PARFAIT, ESTABLISH, and WINS@HI, funded in part by the European Union's Horizon 2020 research and innovation program under grant agreement no. 787002 (SAFECARE) and No. 813278 (A-WEAR).

References

Al Ameen, M., & Kwak, K. S. (2011). Social issues in wireless sensor networks with healthcare perspective. *International Arab Journal of Information Technology, 8*(1), 52–58.

Aleksandrova, M. (2019). *The impact of edge computing on IoT: The main benefits and real-life use cases* [online]. Available at: https://dzone.com/articles/the-impact-of-edge-computing-on-iot-the-main-benef. Accessed 8 Mar 2019.

Ammar, M., Russello, G., & Crispo, B. (2018). Internet of things: A survey on the security of IoT frameworks. *Journal of Information Security and Applications, 38*, 8–27.

Anliker, U., Ward, J., Lukowicz, P., Troster, G., Dolveck, F., Baer, M., Keita, F., Schenker, E., Catarsi, F., Coluccini, L., Belardinelli, A., Shklarski, D., Alon, M., Hirt, E., Schmid, R., & Vuskovic, M. (2004). AMON: A wearable multiparameter medical monitoring and alert system. *IEEE Transactions on Information Technology in Biomedicine, 8*(4), 415–427.

Anon. (2019). *Gas sensor developer kits* [online]. Available at: https://www.spec-sensors.com/product-category/gas-sensor-developer-kits. Accessed 15 Mar 2019.

Anwar, M., Joshi, J., & Tan, J. (2015). Anytime, anywhere access to secure, privacy-aware healthcare services: Issues, approaches and challenges. *Health Policy and Technology, 4*(4), 299–311.

Asada, H. H., Shaltis, P., Reisner, A., Rhee, S., & Hutchinson, R. C. (2003). Mobile monitoring with wearable photoplethysmographic biosensors. *IEEE Engineering in Medicine and Biology Magazine, 22*(3), 28–40.

Bacis, E., De Capitani di Vimercati, S., Foresti, S., Paraboschi, S., Rosa, M., & Samarati, P. (2016). Mix&Slice: Efficient access revocation in the cloud. In *Proceedings of the 2016 ACM SIGSAC Conference on Computer and Communications Security* (pp. 217–228). New York: ACM.

Boesen, P. V., & Dragicevic, D. (2018). *Wireless earpieces utilizing a mesh network.* U.S. Patent Application 15/905, 322.

Brookson, C., Cadzow, S., Eckmaier, R., Eschweiler, J., Gerber, B., Guarino, A., Rannenberg, K., Shamah, J., & Gorniak, S. (2016). *Definition of cybersecurity-gaps and overlaps in standardisation.* Heraklion: ENISA.

Bryk, A. (2018). *Cloud computing: A new vector for cyber attacks* [online]. Available at: https://www.apriorit.com/dev-blog/523-cloud-computing-cyber-attacks. Accessed 8 Jan 2019.

Burleson, W., Clark, S. S., Ransford, B., & Fu, K. (2012). Design challenges for secure implantable medical devices. In *Proceedings of the 49th Annual Design Automation Conference* (pp. 12–17). New York: ACM. https://doi.org/10.1145/2228360.2228364.

Cyr, B., Horn, W., Miao, D., & Specter, M. (2014). *Security analysis of wearable fitness devices (fitbit)* (Vol. 1). Cambridge, MA: Massachusetts Institute of Technology.

Dantu, R., Dissanayake, I., & Nerur, S. (2019). *Exploratory analysis of internet of things (IoT) in healthcare: A topic modeling approach.* Proceedings of the 52nd Hawaii International Conference on System Sciences.

Dey, N., Ashour, A. S., Shi, F., Fong, S. J., & Tavares, J. M. R. (2018). Medical cyber-physical systems: A survey. *Journal of Medical Systems, 42*(4), 74.

Dr. Hempel Digital Health Network. (2019). *Cybersecurity for internet of medical things | A big challenge for healthcare innovators* [online]. Available at: https://www.dr-hempel-network.com/digital-health-technolgy/cybersecurity-for-internet-of-medical-things. Accessed 1 Mar 2019.

Endler, M., Silva, A., & Cruz, R. A. (2017). An approach for secure edge computing in the Internet of Things. In *2017 1st Cyber Security in Networking Conference (CSNet)* (pp. 1–8). Piscataway: IEEE.

ENISA. (2017). *Baseline security recommendations for IoT* [online]. Available at: https://www.enisa.europa.eu/publications/baseline-security-recommendations-for-iot. Accessed 9 Feb 2019.

Farahani, B., Firouzi, F., Chang, V., Badaroglu, M., Constant, N., & Mankodiya, K. (2018). Towards fog-driven IoT eHealth: Promises and challenges of IoT in medicine and healthcare. *Future Generation Computer Systems, 78*, 659–676.

Haghi, M., Thurow, K., & Stoll, R. (2017). Wearable devices in medical internet of things: Scientific research and commercially available devices. *Healthcare Informatics Research, 23*(1), 4–15.

Hiremath, S., Yang, G., & Mankodiya, K. (2014). *Wearable internet of things: Concept, architectural components and promises for person-centered healthcare*. Proceedings of the 4th International Conference on Wireless Mobile Communication and Healthcare – "Transforming healthcare through innovations in mobile and wireless technologies".

Indiegogo. (2019). *Hicon smartwristband with social network icons* [online]. Available at: https://www.indiegogo.com/projects/hicon-smartwristband-with-social-network-icons. Accessed 15 Jun 2019.

Koydemir, H., & Ozcan, A. (2018). Wearable and implantable sensors for biomedical applications. *Annual Review of Analytical Chemistry, 11*(1), 127–146.

Langone, M., Setola, R., & Lopez, J. (2017). Cybersecurity of wearable devices: An experimental analysis and a vulnerability assessment method. In *2017 IEEE 41st Annual Computer Software and Applications Conference (COMPSAC)* (Vol. 2, pp. 304–309). Piscataway: IEEE.

Lin, J., Yu, W., Zhang, N., Yang, X., Zhang, H., & Zhao, W. (2017). A survey on internet of things: Architecture, enabling technologies, security and privacy, and applications. *IEEE Internet of Things Journal, 4*(5), 1125–1142.

Lounis, A., Hadjidj, A., Bouabdallah, A., & Challal, Y. (2016). Healing on the cloud: Secure cloud architecture for medical wireless sensor networks. *Future Generation Computer Systems, 55*, 266–277.

Managed Care. (2016). *Medical device market to hit $133 billion by 2016* [online]. Available at: https://www.managedcaremag.com/archives/2014/8/medical-device-market-hit-133-billion-2016. Accessed 9 Feb 2019.

Mankodiya, K., Hassan, Y. A., Vogt, S., Gehring, H., & Hofmann, U. G. (2010). *Wearable ECG module for long-term recordings using a smartphone processor*. Proceedings of the 5th International Workshop on Ubiquitous Health and Wellness, Copenhagen, Denmark (Vol. 2629).

Mansfield-Devine, S. (2016). Ransomware: Taking businesses hostage. *Network Security, 2016*(10), 8–17.

Martin, G., Martin, P., Hankin, C., Darzi, A., & Kinross, J. (2017). Cybersecurity and healthcare: How safe are we? *BMJ, 358*, j3179.

Mathie, M., Coster, A., Lovell, N., & Celler, B. (2003). Detection of daily physical activities using a triaxial accelerometer. *Medical & Biological Engineering & Computing, 41*(3), 296–301.

Meingast, M., Roosta, T., & Sastry, S. (2006). Security and privacy issues with health care information technology. In *2006 International Conference of the IEEE Engineering in Medicine and Biology Society* (pp. 5453–5458). Piscataway: IEEE.

Monisha, K., & Babu, M. R. (2018). A novel framework for healthcare monitoring system through. In *Internet of things and personalized healthcare systems*. Puchong, Singapore: Springer.

Monisha, K., & Babu, M. R. (2019). A novel framework for healthcare monitoring system through cyber-physical system. In *Internet of things and personalized healthcare systems* (pp. 21–36). Puchong, Singapore: Springer.

Muncaster, P. (2016). NHS Trust suspends operations after major cyber incident. *Infosecurity* [online]. Available at: http://www.infosecurity-magazine.com/news/nhs-trust-suspends-operations. Accessed 2 Mar 2019.

Palmer, D. (2018). *Edge computing: The cybersecurity risks you must consider* [online]. Available at: https://www.zdnet.com/article/edge-computing-the-cyber-security-risks-you-must-consider. Accessed 2 Feb 2019.
Pan, J., & Yang, Z. (2018). Cybersecurity challenges and opportunities in the new edge computing + IoT world. In *Proceedings of the 2018 ACM International Workshop on Security in Software Defined Networks & Network Function Virtualization* (pp. 29–32). New York: ACM.
Pew Research Center. (2019). *Pew Internet and American life project* [online]. Available at: http://www.pewinternet.org/2011/02/01/health-topics-4/. Accessed 9 Feb 2019.
Piggin, R. (2017). Cybersecurity of medical devices: Addressing patient safety and the security of patient health information. *BSI Group, Macquarie Park, Australia, White Paper.*
PriceWaterhouseCoopers. (2016). *Insurance 2020: Reaping the dividends of cyber resilience* [online]. Available at: http://www.pwc.com/gx/en/industries/financial-services/insurance/publications/insurance-2020-cyber.html. Accessed 4 Mar 2019.
Qureshi, F., & Krishnan, S. (2018). Wearable hardware design for the internet of medical things (IoMT). *Sensors (Basel, Switzerland), 18*(11), 3812.
Roetenberg, D. (2006). *Inertial and magnetic sensing of human motion*. University of Twente.
Seals, T. (2019). *Fitbit vulnerabilities expose wearer data* [online]. Available at: https://www.infosecurity-magazine.com/news/fitbit-vulnerabilities-expose. Accessed 5 Mar 2019.
Seoane, F., Ferreira, J., Alvarez, L., Buendia, R., Ayllón, D., Llerena, C., & Gil-Pita, R. (2013). Sensorized garments and textrode-enabled measurement instrumentation for ambulatory assessment of the autonomic nervous system response in the ATREC project. *Sensors, 13*(7), 8997–9015.
Stanley, N., & Coderre, M. (2016). *An introduction to medical device cyber security: A European perspective.*
Sun, W., Cai, Z., Li, Y., Liu, F., Fang, S., & Wang, G. (2018). Security and privacy in the medical internet of things: A review. *Security and Communication Networks, 2018*, 1–9.
Xu, S., Zhang, Y., Jia, L., Mathewson, K., Jang, K., Kim, J., Fu, H., Huang, X., Chava, P., Wang, R., Bhole, S., Wang, L., Na, Y., Guan, Y., Flavin, M., Han, Z., Huang, Y., & Rogers, J. (2014). Soft microfluidic assemblies of sensors, circuits, and radios for the skin. *Science, 344*(6179), 70–74.
Yang, G., Xie, L., Mantysalo, M., Zhou, X., Pang, Z., Xu, L., Kao-Walter, S., Chen, Q., & Zheng, L. (2014). A health-IoT platform based on the integration of intelligent packaging, unobtrusive bio-sensor, and intelligent medicine box. *IEEE Transactions on Industrial Informatics, 10*(4), 2180–2191.
Zhao, Y. (2016). Identity-concealed authenticated encryption and key exchange. In *Proceedings of the 2016 ACM SIGSAC Conference on Computer and Communications Security* (pp. 1464–1479). New York: ACM.

Index

S. K. Shandilya et al. (eds.), *Advances in Cyber Security Analytics and Decision Systems*, EAI/Springer Innovations in Communication and Computing,
https://doi.org/10.1007/978-3-030-19353-9

MIX
Papier aus verantwortungsvollen Quellen
Paper from responsible sources
FSC® C105338

If you have any concerns about our products,
you can contact us on
ProductSafety@springernature.com

In case Publisher is established outside the EU,
the EU authorized representative is:
Springer Nature Customer Service Center GmbH
Europaplatz 3, 69115 Heidelberg, Germany

Printed by Libri Plureos GmbH
in Hamburg, Germany